Engineering Asset Management Review

Volume 3

Engineering Asset Management Review (EAMR) is published under the auspices of the International Society of Engineering Asset Management (ISEAM). Engineering Asset Management focuses on life-cycle management of the physical assets required by a private or public firm, for the purpose of making products, and/or for providing services in a manner that satisfies various business performance rationales. In exploring the wide ranging issues involved in the management of engineered assets constituting our built environment, EAMR takes a broad view of the inter- and multi-disciplinary approach which combines science, engineering, and technology principles with human behavior and business practice.

Adolfo Crespo Márquez • Turuna S. Seecharan
Georges Abdul-Nour • Joe Amadi-Echendu
Editors

Advances in Asset Management: Strategies, Technologies, and Industry Applications

Springer

Editors
Adolfo Crespo Márquez
Department of Industrial Management
University of Seville
Seville, Spain

Georges Abdul-Nour
Université du Québec à Trois-Rivière
Trois-Rivières, QC, Canada

Turuna S. Seecharan
Department of Mechanical & Industrial Engineering
University of Minnesota, Duluth
Duluth, MN, USA

Joe Amadi-Echendu
Graduate School of Technology Management
University of Pretoria
Hatfield, Pretoria, Gauteng, South Africa

ISSN 2190-7846 ISSN 2190-7854 (electronic)
Engineering Asset Management Review
ISBN 978-3-031-52390-8 ISBN 978-3-031-52391-5 (eBook)
https://doi.org/10.1007/978-3-031-52391-5

This Springer imprint is published by the registered company Springer Nature Switzerland AG
The registered company address is: Gewerbestrasse 11, 6330 Cham, Switzerland

Paper in this product is recyclable.

Preface

Welcome to the third volume of the Springer EAMR Series – a prestigious collection under the auspices of the International Society of Engineering Asset Management (ISEAM). This volume, titled *Advances in Asset Management: Strategies, Technologies, and Industry Applications*, is a testament to the ongoing commitment of ISEAM and Springer to provide a platform for the dissemination of cutting-edge research in the field of engineering asset management. This compilation, meticulously elaborated, brings together selected papers that initially appeared in the proceedings of the 16th World Congress on Engineering Asset Management (WCEAM). Recognizing the value and significance of these contributions, the authors were invited to extend and enhance their papers, resulting in the comprehensive and enriched content you hold in your hands. We express our sincere gratitude to the authors for their dedication and support to ISEAM and the Springer EAMR Series. Their commitment has played a pivotal role in the continued success of this collaborative endeavor. As we embark on this journey through the realms of asset management, we invite readers to explore the nuanced discussions, innovative strategies, and practical insights presented in this volume, each contributing to the overarching goal of advancing the field and fostering excellence in asset management practices worldwide.

The content of the book is divided in the following four parts:

Part I: Risk Management and Qualitative Analysis

In the opening part, we lay the foundation with an exploration of risk management principles and qualitative analysis. The Risk Qualitative Criticality Matrix (RQCM) sets the stage, introducing readers to the fundamental concepts of risk assessment. Additionally, we delve into the critical factors influencing the quality of network services in emerging telecom markets, offering valuable insights into the complex interplay of risks in dynamic operating environments.

Part II: Technology and Innovation in Asset Management

Part II focuses on the integration of technology and innovation in asset management. "A Conceptual Implementation Process for Smart Maintenance Technologies" provides a roadmap for leveraging cutting-edge solutions. Complementing this, "A Framework for Assessing Emerging Technology Risks in Industrial Asset" explores the challenges and opportunities presented by emerging technologies in industrial settings.

Part III: Asset Health and Maintenance Strategies

This section delves into strategies for maintaining asset health. "Challenges on an Asset Health Index Calculation" guides readers through the complexities of calculating and maintaining an Asset Health Index, while "Determination of the Exact Economic Time for the Component Replacement Using Condition-Based Maintenance" navigates the economic considerations of optimal replacement timing. The general bases for hierarchy definition for digital assets in the railway context are also explored.

Part IV: Industry-Specific Asset Management and Other Considerations

The final part consolidates industry-specific insights and additional considerations. The audit models take center stage, with "Audit Models for Asset Management, Maintenance and Reliability Processes: A Case Study Applied to the Desalination Plant" and "Audit Model for Asset Management, Maintenance and Reliability Processes: A Case Study Applied to Pulp Mill Sector". These studies provide practical applications of audit models, offering real-world examples of their effectiveness. Additionally, "The Role of Eco-Driving and Wearable Sensors in Industry 4.0" expands the discussion to the role of sustainability and emerging technologies in the future of asset management.

This book serves as a comprehensive guide for professionals, researchers, and students seeking a deep understanding of contemporary asset management practices. Each part unfolds a new dimension, collectively contributing to the ongoing dialogue on how organizations can optimize their assets in an ever-evolving landscape. We invite you to explore the rich tapestry of ideas and insights presented in *Advances in Asset Management: Strategies, Technologies, and Industry Applications*.

Seville, Spain Adolfo Crespo Márquez

Duluth, MN, USA Turuna S. Seecharan

Trois-Rivières, QC, Canada Georges Abdoul-Nour

Hatfield, Pretoria, Gauteng, South Africa Joe Amadi-Echendu

Contents

About the Editors

Adolfo Crespo Márquez is currently Full Professor at the Department of Industrial Management, School of Engineering of the University of Seville. He holds a PhD with Honors in Industrial Engineering from this same university. He is included in the Stanford University's list of world top-scientists in 2023 and he received the Spanish Maintenance Award from the Spanish Maintenance Association (AEM) in 2020. With research works published in many top journals, he is the author of 8 books and editor of 6 books with Springer, Aenor, IGI Global and INGEMAN, about maintenance, warranty, supply chain and assets management. Prof. Crespo is Fellow Director of ISEAM, Editor-in-Chief of EAMR Springer Book Series and led the Spanish Research Network on Assets Management and the Spanish Committee for Maintenance Standardization (1995–2003). He also leads the SIM research group related to smart maintenance systems and has extensively participated in many engineering and consulting projects for different companies, for the Spanish Departments of Defense, Science and Education as well as for the European Commission (IPTS). He is the President of INGEMAN (a National Association for the Development of Maintenance Engineering in Spain) since 2002.

Georges Abdul-Nour is an Emeritus Professor at the Department of Industrial engineering, Université du Québec à Trois-Rivières, Canada. He received his Bachelor and Master Degrees in IE from the Université de Moncton, NB, Canada. He completed his PhD at Texas Tech University; Lubbock Texas, U.S.A. Dr. Abdul-Nour authored and co-authored 10 books and book chapters, and more than 450 refereed papers in international journals and conferences proceeding, and technical reports. His research interest is in fields such as Intelligent Asset Management, Complexes Systems Resilience, Reliability and Maintenance, Intelligent Manufacturing, I4.0 and I5.0, Lean-Six Sigma and JIT (TPS). He is a member of the Editorial Board of the *International Journal of Production Research* (IJPR), and Fellow and board member of ISEAM. Professor Abdul-Nour is a registered member of the Order of Engineers of Quebec (OIQ). He served as Chair of Engineering School and of Research Institute on Small and Medium Enterprises (SME),

INRPME at UQTR. He is an elected member of the Academy of Industrial Engineering of Texas Tech. Dr. Abdul-Nour hold the HQ/NSERG Research Chair in Asset Management.

Turuna S. Seecharan obtained her MSc and PhD degrees in Systems Design Engineering from the University of Waterloo, Canada, specializing in "Robust Design Optimization using Metamodels". She is a licensed professional engineer with the Professional Engineers of Ontario and a board member of the International Society of Engineering Asset Management. Presently, she is an Assistant Professor in the Department of Mechanical and Industrial Engineering at the University of Minnesota Duluth. Her current research interests are Gamification and the role of Human Factors in Asset Management.

Joe Amadi-Echendu is an Emeritus Professor at the Graduate School of Technology Management, the Faculty of Engineering, Built Environment, and Information Technology at the University of Pretoria. He specializes in the management of Engineering and Technology and has been consulted on a variety of issues regarding technology commercialization and asset management. Joe continues to apply knowledge and expertise in public and private sector industries. He has co-edited 7 books, sole-authored a book and authored/co-authored over 150 articles in various technical publications worldwide. Joe obtained the chartered engineer status in the UK, and he is a registered professional engineer as well as a NRF rated researcher in South Africa. Joe recently served as an experienced researcher for the European Union Horizon 2020 MSCA RISE Project 645733 (2014–2018) on the Sustainable Design and Management of Industrial Assets through Total Value and Cost of Ownership. Joe is a Founding Fellow, Director and current Chair of the Board of Directors of the International Society of Engineering Asset Management (www.iseam.org), as well as a member of the Board of International Association for the Management of Technology (www.iamot.net).

Part I
Risk Management and Qualitative Analysis

RQCM: Risk Qualitative Criticality Matrix. Case Study: Ophthalmic Lens Production Systems in Costa Rica

Carlos Parra, Juan Rodríguez, Adolfo Crespo Márquez, Vicente González-Prida, Pablo Viveros, Fredy Kristjanpoller, and Jorge Parra

Abstract The use of prioritization analysis techniques allows identifying the level of criticality of physical assets and helps to manage resources: human, economic and technological in a more efficient way. In other words, the process of criticality analysis helps to determine the importance and consequences of the failures of productive equipment in the operational context in which they perform. This article explains the basic theoretical aspects of the equipment prioritization analysis process based on risk matrices (failure frequency and consequences); and the development of the model named Risk Qualitative Criticality Matrix (RQCM). Finally, are presented and analysed the results of a case of application of the RQCM in the sector of ophthalmic lenses (new factory built in Costa Rica – PRATS Laboratory).

1 Introduction

Taking as reference the 8-phase of Maintenance Management Model (MMM) (see Fig. 1), this section describes the hierarchical and criticality techniques and is related to phase 2 of the MMM (Crespo, 2007; Parra & Crespo, 2015). The techniques of criticality analysis are tools that allow identifying and hierarchy for their importance the assets of an installation on which it is worth directing resources

C. Parra (✉) · P. Viveros · F. Kristjanpoller
Universidad Técnica Federico Santa María, Valparaíso, Chile
e-mail: carlos.parram@usm.cl

J. Rodríguez
Instituto Tecnológico de Costa Rica, Cartago, Costa Rica

A. Crespo Márquez · V. González-Prida
Universidad de Sevilla, Seville, España

J. Parra
Universidad Tecnológica de Panamá, Panama City, Panamá

A. Crespo Márquez et al. (eds.), *Advances in Asset Management: Strategies, Technologies, and Industry Applications*, Engineering Asset Management Review 3, https://doi.org/10.1007/978-3-031-52391-5_1

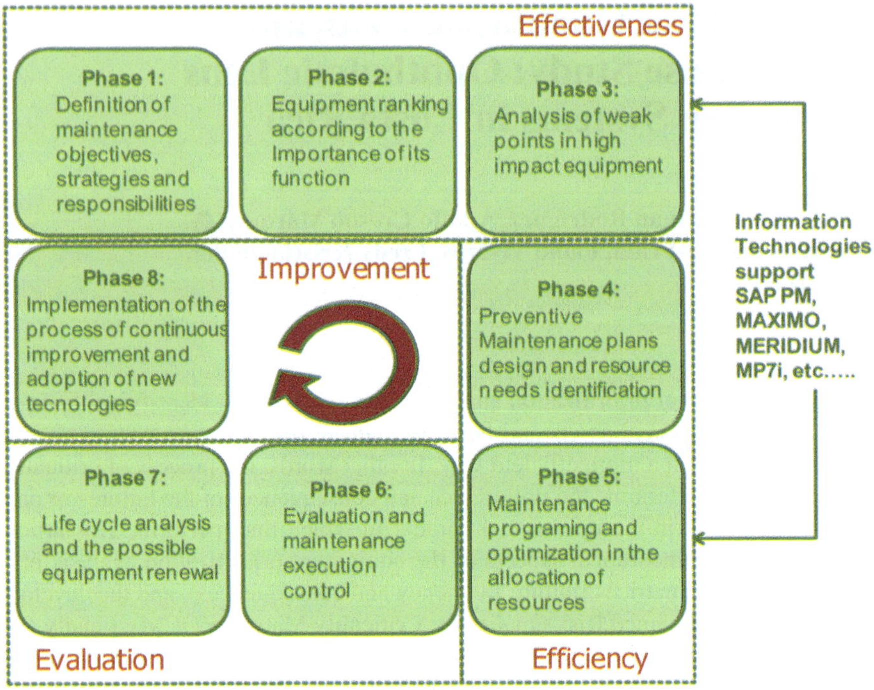

Fig. 1 Maintenance management process model. Source: Crespo Márquez et al., 2009

(human, economic and technological). In other words, the process of criticality analysis (also known as risk assessment) helps to determine the importance and consequences of potential failure events of production assets within the operational context in which they work.

Criticality analysis is a great tool for identifying the priority of maintenance tasks. A good way to look at it is that maintenance task priority should be established by the risk level that comes with not performing that task. Coincidently, this level of risk associated with not doing a particular maintenance task is determined by the consequences of the potential failure that could happen if the task is not completed and the likelihood of that failure occurring if the task is not done at a predetermined time. Once you have your criticality ratings, a criticality analysis can help you choose a proper risk mitigation strategy that you can apply to each asset. The term "critical" and the definition of criticality can have different interpretations depending on the objective that is being treated by hierarchy (Crespo Márquez et al., 2009; Parra et al., 2021a). The objective of a critical analysis is to establish a method that serves as an instrument of aid in determining the hierarchy of processes, systems, and equipment of a complex production process, allowing subdividing the elements in sections that can be handled in a controlled and auditable manner. From this perspective there is a great diversity of possible criteria that allow us to evaluate the criticality of an asset of production. Prioritization reasons may vary according to the opportunities and needs of the organization (Crespo, 2007). Below are some common criteria to be used within the hierarchy processes:

- Operational flexibility (availability of alternating function or backup)
- Effect in operational continuity /production capacity
- Effect in product quality
- Effect on health, safety and environment
- Costs of shutdown and maintenance
- Failure frequency/reliability
- Operating conditions (temperature, pressure, fluid, flow, speed) Flexibility/ Accessibility for inspection & Maintenance
- Requirements/Resource availability for inspection and maintenance
- Availability of spare parts

The risk assessment process starts by first identifying risk events. In turn, these risk events have two dimensions (Fig. 2):

- The consequence of an event
- The likelihood of an event

The overall level of risk is determined by the combination of these dimensions, frequently visualized in a risk matrix. We can consider risk to be the combination of the severity of consequences of an event, and the probability or likelihood of that event occurring. In other words, risk applies to an event – not to a physical item (such as an item of equipment). If we consider that Equipment criticality is the same as equipment failure risk, then we had better be clear about what the failure event(s) are that we are assessing.

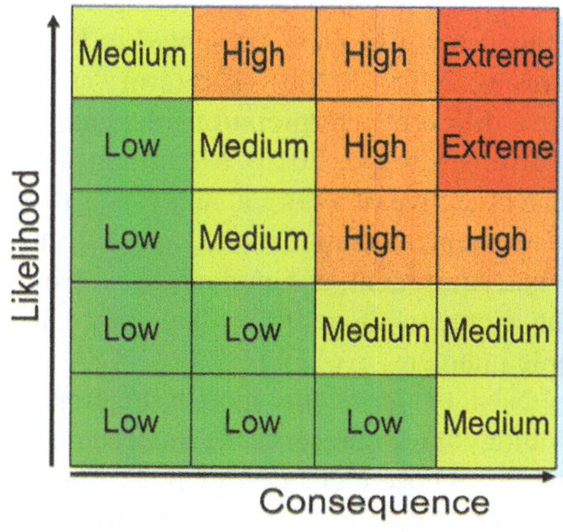

Fig. 2 Qualitative Risk Matrix (QRM). Source: Parra et al., 2020a, b

Even the recently published ISO 55000 standard for Asset Management does not define equipment criticality – although it does define a critical asset as being an "asset having potential to significantly impact on the achievement of the organization's objectives". ISO 55002 suggests that a "a risk ranking process can be used to determine which assets have a significant potential to impact on the achievement of the asset management objectives, i.e., which are the critical assets". However, once again, assessing risk implies having to assess the likelihood of an event, which in turn means that we need to be clear about exactly which events are being assessed, and how the probability and consequences associated with multiple events on an equipment item are to be rolled up to an overall failure risk associated with that equipment (Parra et al., 2020a, b). Furthermore, if we accept that Equipment Criticality is somehow derived from equipment failure risks, it is not clear whether we are intended to assess the unmitigated or mitigated risks associated with each of these failures. In other words, are we supposed to assess the risks assuming that we do not have any controls in place to minimize the likelihood or consequences of those failure events (or that they are not effective) or are we expected to assume that the controls that we currently have in place are effective when assessing equipment failure risks? (Crespo Márquez et al., 2009; Parra et al., 2021a).

Qualitative hierarchizing techniques based on risk analysis are tools that can be used to determine the criticality of industrial business assets. These techniques allow us to evaluate and know the level of importance of industrial assets considered two factors: frequency and consequences of failures and help those involved in decision-making processes effectively guide resources: human, economic and technological in the areas of maintenance, operations, logistics, quality, safety, environment, etc. In other words, the process of criticality risk analysis (CRA), helps determine the importance of assets according to the consequences caused by failure events in the operational context in which they work (Parra et al., 2021a; Neurohr et al., 2021). The following article, takes as a specific reference, the risk-based ranking proposal developed in Phase 2 of the MMM (Maintenance Management Model), presented in Fig. 1 (Crespo Márquez et al., 2009, Parra & Crespo, 2015, 2020b). The term "critical" and the definition of criticality may have different interpretations depending on the goal that is trying to hierarchize (Parra & Crespo, 2015, 2019; González-Prida et al., 2012; Parra et al., 2021b). The objective of a critical analysis is to establish a method that serves as a generic instrument in maintenance; and help to determine the hierarchy of plants, systems, equipment, components, etc., from a complex production process, allowing the elements in sections that they can be handled in a controlled and auditable manner.

Next, two criticality models are presented: QRM: Qualitative Risk Matrix, methodology that represents the antecedent to the RQCM: Risk Qualitative Criticality Matrix, whose methodology is the one that will to be used in the case study of this chapter: Ophthalmic lens production systems in Costa Rica, both based on the risk assessment process and used to identify critical systems (Parra et al., 2020a, b, 2021a).

2 Criticality Model QRM: Qualitative Risk Matrix

Risk assessment is a fundamental process of the oil refining industry that allows to allocate priorities related to mitigation plans and the selection of maintenance strategies. In industrial risk assessments, they combine the probability/frequency of a failure event with the impact that the fault event would cause (Parra et al., 2021b). The decision-making process behind the determination of the criticality of assets requires a hierarchical structure and the application of some mathematical models that allow weights and priorities of assets to be evaluated. In this case study, the steps to be followed to design the risk-based criticality model would be the following (Parra et al., 2020a, b, 2021b):

1. Define a scope and purpose for the criticality analysis based on the risk model
2. Define the level of detail of the analysis (Taxonomy – ISO 14224 Standard Reference)
3. Establish criteria of importance of the risk model: ranges of failure frequencies and the factors of consequences to be evaluated (aligned with the business objectives) within the risk model
4. Select or develop a risk assessment method that allows hierarchy systems

The criticality model taken as a reference in this section, is called Qualitative Risk Matrix – QRM, originally designed for the off-shore production assets of the Magallanes Oil Production ENAP SIPETROL and adapted to refinery plants (Parra & Crespo, 2015; Parra et al., 2021b). The proposed model is based on the estimation of the risk factor and adjusted to the needs of the catalytic cracking unit. The general expressions for the evaluation of the criticality model based on the risk factor are presented below:

Criticality Index

$$\mathrm{CI} = \mathrm{FF} \times \mathrm{C} \tag{1}$$

Where:
CI: Criticality Index
FF: is the frequency factor or the number of faults in a given period (failures/year)
C: is the general factor of consequences of the failure, this factor is divided into different subcategories:

Consequences

$$\mathrm{C} = (\mathrm{OI} \times \mathrm{OF}) + \mathrm{MC} + \mathrm{HSEI} \tag{2}$$

Where:
OI: Operational Impact
OF: Operational Flexibility
MC: Cost of Corrective Maintenance
HSEI: Health, Safety, and Environment Impact

It is important to mention that the factors included in the QRM (frequencies and consequences) were proposed and endorsed by the management of the refinery under study. For the subcategories of consequences (C = (OI × OF) + MC + HSEI)), the percentage (%) of importance assigned to each factor of the consequences of the matrix are aligned with the objectives defined by the management of the Oil Refinery complex and were approved by the business owners. The importance weights of the consequences of the failures are presented below:

OI × OF = accounts for 80% of the total weight of the consequences (40/50 = 80%)
MC = accounts for 4% of the total weight of the consequences (2/50 = 4%)
SEI = accounts for 16% of the total weight of the consequences (8/50 = 16%)

Regarding the definition of the factors: frequency of failures (F) and consequences C = (OI × OF) + MC + HSEI)), the natural work team, established the intervals and the measuring scales to classify the different assets in the developed risk matrix (see Fig. 3 and Tables 1, 2, 3, 4 and 5).

Regarding the development of the risk matrix, it has the following configuration:

- Vertical axis (failure frequency): 4 rows, maximum value 4 points (scale: 1–4)
- Horizontal axis (failure consequences): 5 columns, maximum value: 50 points (scale of 1 to 50)

The values of the attributes for subcategories of consequences in the QRM are assigned so that they are aligned with the business objectives, and the approval of the management of the installation for its application in the analysis is required. Below are the weighting factors for frequency (F) and consequences (C) to be used for the evaluation of criticality in the risk matrix (Parra & Crespo, 2015).

The results of the evaluation of the above factors will allow defining the criticality of the assets evaluated in the QRM (see Fig. 3). The vertical axis is formed by 4

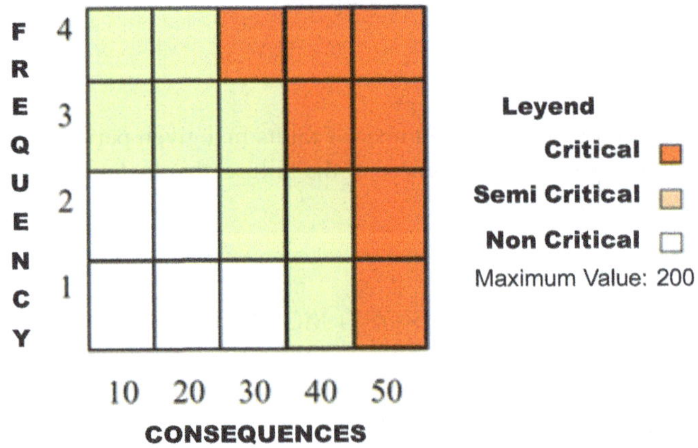

Fig. 3 Qualitative Risk Matrix (QRM). Source: Parra et al., 2020a, b

Table 1 Classification and scale (FF): Failure frequencies

Failure frequency F	Failures per year	Value
Poor	> 4	4
Average	3–4	3
Good	1–2	2
Excellent	< 1	1

Source: Own elaboration

Table 2 Classification and scale OI: Operational impact

Operational Impact (OI)	Consequences	Scale
Extremely high	Overall plant total loss of production	10
High	Loss of production in a main process	8
Average	Partial loss in a production process	5
Low	Minor losses in a production process	3
No impact	No impact in production	1

Source: Own elaboration

Table 3 Classification and scale OF: Operational flexibility

Operational flexibility (OF)	Consequences	Scale
High	No backup / sparc cquipment nor alternate operational procedure available	4
Average	Share backup / spare equipment available	2
Low	Backup / spare equipment available	1

Source: Own elaboration

Table 4 Classification and MC scale: Cost of corrective maintenance

Maintenance Costs (MC)	Consequences	Scale
High	C ≥ 20,000 US$	2
Low	C < 20,000 US$	1

Source: Own elaboration

Table 5 HSEI classification and scale: health, safety and environment impact

SEI	Consequences	Scale
Extremely high	External and internal catastrophic environmental impact/loss of lives (death) and/or physical damage with permanent disabling injuries to people, requiring notification to public organizations	8
Very high	Irreversible environmental impact/physical damage with temporary disabling injuries to people with loss of work time/ severe damage in facilities	6
Average	Reversible environmental impact/physical damage with injuries that require immediate attention to people by health services, but without causing disability or loss of work time	4
Low	Minor incidents and accidents (recoverable)	2
No impact	No effect in people, environment, and facilities	1

Source: Own elaboration

frequency levels of failures, while the horizontal axis is formed by five levels of failure consequences. To define the level of equipment criticality to be evaluated, a group of experts analyze the frequency factors and consequence of the failures and allocate rating to each factor according to the values shown in the previous tables. In summary, the definition of criticality is detailed below: First, the value assigned to the failure frequency (FF) is selected and the vertical position is defined in the matrix of the evaluated equipment, then the values are assigned to the 4 factors that make up the consequences of failures (C = (OI x OF) + MC + HSEI)), the result of Eq. 2 represents the value that will define the horizontal position of the equipment evaluated in the criticality matrix. Subsequently, the frequency value and the value obtained on the side of the consequences are taken and said values are intercepted in the risk matrix (Fig. 2). In this way, the level of criticality of the equipment evaluated in the risk matrix is obtained (the risk matrix is divided into three regions representing three levels of criticality):

NC: Non Critical
SC: Semi Critical
C: Critical

Here is a basic example of the use of the matrix:

- Equipment to evaluate: M101

FF engine: 3 average (2–4 faults per year)

OI: 5 averages (partial loss of a production process)
OF: 2 averages (shared backup function)
MC: 2 High (C ≥ 20,000 US $)
HSEI: 4 averages (Reversible Environmental Impact/Minor Physical Damage)
Final position (values to intercept in the risk matrix):
Failure frequency (FF): 3
Consequence factor: = (OI x OF) + MC + HSEI = (5 x 2) + 2 + 4 = 16
Position in the criticality matrix: 3 vertical axis, 16 horizontal axis = SC (semi critical), (see Fig. 3)

3 Criticality Model RQCM: Risk Qualitative Criticality Matrix

Risk assessment techniques can be used to prioritize equipment/assets and align maintenance actions to key business objectives (Parra et al., 2021a; Li & Wright, 2019). When carrying out, it ensures that maintenance actions are effective from the point of view of the main costs associated with maintenance and most importantly be efficient to minimize the consequences on safety, environment, production (Parra et al., 2021b; Junietz et al., 2018). In this case: Ophthalmic lens production systems in Costa Rica, the steps to be followed to design the risk-based criticality model are similar to the

QRM method presented in the previous section and it was adjusted to the needs of the industry of the ophthalmic sector (Parra & Crespo, 2018, 2019; Parra et al., 2021c):

1. Define a scope and purpose for criticality analysis based on the risk model. This will be defined according to maintenance goals aligned to business goals and management.
2. Define the level of detail of the analysis (Taxonomy Reference – ISO 14224 Standard).
3. Importance criteria of the risk model should be established: ranges of fault frequencies (FF) and the consequence factors (C) to be evaluated (aligned with the business objectives) within the selected risk model.
4. Selecting or developing a risk assessment method that allows the systems within the industry or department. The criticality model taken as a reference for this article is called Risk Qualitative Criticality Matrix (RQCM), originally it was designed for the off-shore production assets of the Magallanes area, ENAP Sipetrol, (ENAP Sipetrol, 2015, 2016; Crespo, 2007; Parra & Crespo, 2020a, b; Viveros-Gunckel et al., 2020).

For this case, the model of the qualitative criticality risk: RQCM, is used in the ophthalmic lens company, hoping to be the door to future risk assessments, directing the maintenance management of PRATS Laboratory towards continuous improvement with the proposed recommendations and the lessons learned. The RQCM (Parra et al., 2020a; Chiu et al., 2017), is a simple analysis process, which is supported in the concept of risk: frequency of a failure by the consequences, the expression used for hierarchy systems from the RQCM model is:

Total Criticality Risk

$$TCR = FF \times C \tag{3}$$

Where,

- Total Criticality Risk
- Failure frequency (failure range in a certain time (failures/year)
- Consequences of failure events

Where the value of the consequences (C) is obtained from the following expression:

Estimation of Consequences

$$C = (PI \times OF) + MC + HSE \tag{4}$$

Where,

- Impact on production factor
- Operational flexibility factor
- Maintenance costs factor
- Health, Safety and Environment factor

The final expression of the TCR prioritization model will be as follows:

Total Criticality Risk

$$\mathrm{TCR} = \mathrm{FF} \times \left((\mathrm{PI} \times \mathrm{OF}) + \mathrm{MC} + \mathrm{HSE} \right) \tag{5}$$

The weighted factors of each criteria are presented below (criteria were adjusted to the needs of the industry of the ophthalmic sector: PRATS Costa Rica laboratory)

- Fault frequency factor (FF) (scale 1–4)

4: Frequent: greater than 2 events per year
3: Average: 1 and 2 events per year
2: Good: Between 0.5 and an event a year
1: Excellent: Less than 0.5 events a year

- Factors of consequences or operational impact (PI) (scale 1–10)

10: Production losses greater than 75%
7: Production losses between 50% and 74%
5: Production losses between 25% and 49%
3: Production losses between 10% and 24%
1: Production losses less than 10%

- Impact by operational flexibility (OF) (scale 1–4)

4: No backup units are available to cover production, long repair times and complicated logistics
2: There are backup units that they can be partially cover the impact of production, average repair times and logistics
1: It has standby, there is no affectation in the process

- Impact on maintenance costs (MC) (scale 1–2)

2: Repair costs, materials and labour exceeding 20,000 dollars
1: Repair costs, materials and labour less than 20,000 dollars

- Impact on Health, Safety and Environment (HSE) (Scale 1–8)

8: High risk of lives losses, serious health damage, higher environmental incident (Catastrophic) that exceeds the allowed limits
6: Average risk of loss of life, important damage to health, environmental incident of difficult restoration
3: Minimum risk of loss of life and health condition (recoverable in the short term) and/or minor environmental incident (controllable), easy-to-contain and repeated leakage
1: There is no risk of loss of life, no health condition, or environmental damage

The selection of the weighted factors is carried out in work meetings with the participation of the different persons involved in the operational context of the asset in study (operations, maintenance, processes, safety and environment). To obtain the level of criticality of each equipment/business system, the total values of each of the

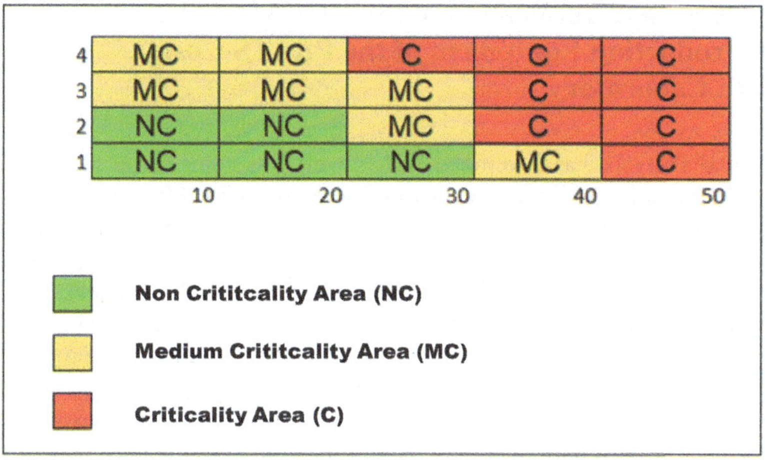

Fig. 4 Equipment criticality matrix. Source: Parra et al., 2021a

main factors are taken: frequency and consequences of the failures and are located in the 4 × 5 criticality matrix (Fig. 4). The failure frequency value is located on the vertical axis and the consequence value is located on the horizontal axis (the final result of the consequence expression is taken: ((PI x OC) + CM + HSE calculated). The criticality matrix shown next, allows systems to be ranked in three areas (see Fig. 4):

- Area of Non-Critical systems (NC)
- Medium Criticality Systems Area (MC)
- Critical Systems Area (C)

The result of the equation is located in the matrix (Fig. 4) to determine which area the equipment under study is located. The maximum value of risk criticality that can be obtained is 200 points distributed in 3 possible levels of hierarchizing systems (critical, semi critical and not critical). With regard to the development of the risk matrix, it has the following configuration:

- Vertical axis (Failure Frequency): 4 rows, maximum value 4 points (scale: 1 to 4).
- Horizontal axis (Failure Consequences): 5 columns, maximum value: 50 points (scale: 1 to 50).

For subcategories of consequences (C = (PI x OF) + MC + HSE)), the percentage of importance assigned to each factor of the consequences, is aligned with the business objectives and they were approved by company management:

- PI × OF = represents 80% of the total weight of the consequences: (40/50 = 80%)
- MC = represents 4% of the total weight of the consequences: (2/50 = 4%)
- HSE = represents 16% of the total weight of the consequences: (8/50 = 16%)

4 Case Study: Application of the RQCM Model in Production Equipment of the PRATS Costa Rica Laboratory

The critical analysis was developed at the level of systems (level 5, according to ISO 14224, see Fig. 5) in the new laboratory of Grupo PRATS. Specifically, were evaluated the 12 more relevant systems that are included as technical locations in the hierarchical structure of the laboratory (Orders Workshop (TE), Treatments (TRA), Control (CT) and Beveling and assembly (BM)) (Rodríguez, 2021). It is important to make note that the application of the RQCM model can be performed at different hierarchical levels as shown in the pyramid in Fig. 5 (plant, processes, systems, equipment, components, etc.), and the results will be specific for the study carried out. For example, the criticality of two similar systems in the same industry may be different since risk factors for both systems may vary or have a different relative importance depending on the characteristics of the operational context in which each system it is operating (Parra et al., 2020b). In this study, the natural work team was composed of five members, including the facilitator and the people of the following departments: technical management, maintenance

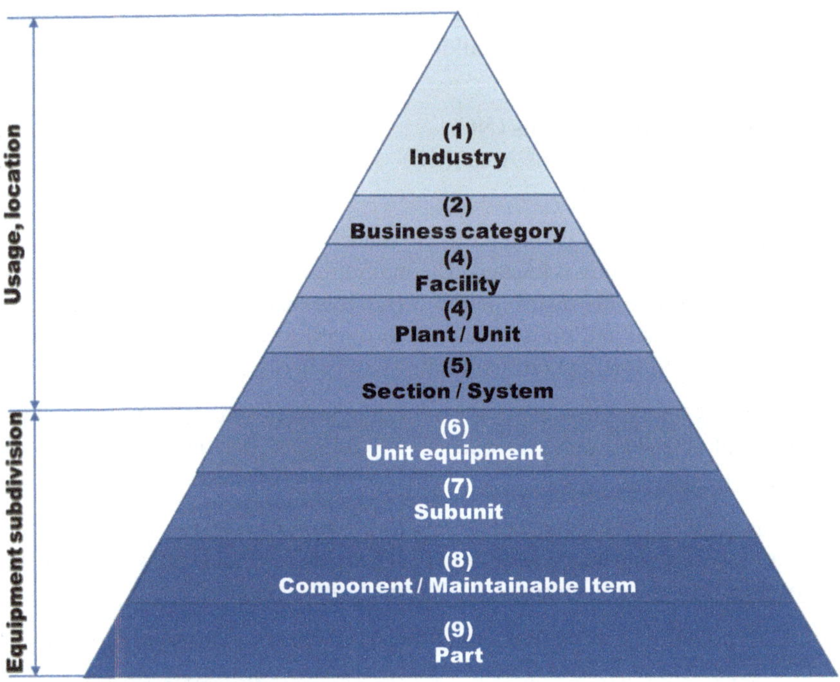

Fig. 5 Proposed taxonomy for the hierarchization of physical assets using the different levels of ISO 14224. Source: Parra et al., 2021a

management, operations management, operators trained in Brazil and Italy with extensive experience of equipment and maintenance.

4.1 Description of the Productive Process and Operational Context

PRATS Laboratory of Costa Rica, specializes in the production of ophthalmic lenses and sunglasses, together with their respective treatments (against scratch, reflection, glare; mirrored, tinted, among others). This laboratory produces 1000 daily lenses between finished and semi-finished (equipment availability required = 88%). This is proposed for a good start of the factory and a competitive insertion in the Costa Rican market. PRATS manufacturing process is divided in the following areas:

- Orders Workshop (TE)
- Treatments (TRA)
- Control (CT)
- Beveling and assembly (BM)

Simplified manufacturing process (see Fig. 6):

1. Choice of the necessary block and tools: the most suitable block is chosen, its material (resins CR-39, MR-8 or polycarbonate) and the molds for the tuning and polishing of each surface.
2. Generation of the anterior surface of the lens: consists of four stages: clamping, generation, tuning and polishing.
3. Intermediate Control: the first surface of the sagitta lens and the thickness are controlled.
4. Generation of the posterior surface of the lens: consists of four stages: clamping, generation, tuning and polishing.
5. Treatment: the lenses are washed and healed in ultrasonic washing machine before entering and the baking white room at approximately 120 °C do the treatment of multilayer (against scratches, reflection, glare; mirrored, tinted, etc.). This is done in clean room laboratory with controlled environment Grade C/ISO 7, air quality PM 2.5 and at 22 °C in the environment.
6. Beveling of the lens to give it the desired shape by the client, this is carried out with the Italian machine of cutting MEI and is a crucial step, it is the final step before mounting the pair of lenses in the desired frame (see Fig. 4, semi-finished lenses).
7. Final control: Controls the quality of the surfaces, the mass and the recipe formulated for the client (Robotic AR).
8. Packaging and storage: Later it is delivered to the messengers for shipping to the customer.

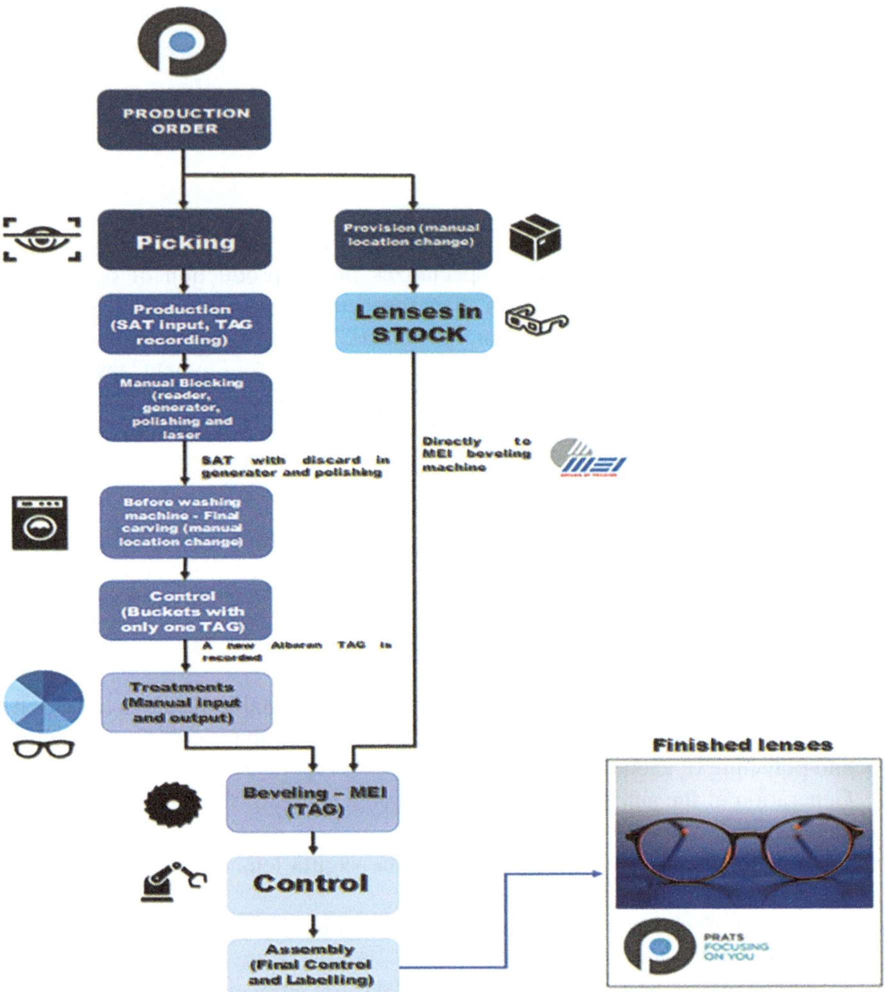

Fig. 6 PRATS productive process flow diagram. Source: Rodríguez, 2021

4.2 *Results and Analysis of the RQCM Application*

The results of the application of the RQCM (Tables 6, 7 and Fig. 7) are summarized, the maximum value of risk criticality that can be obtained is 200 points (TCR: Total Criticality Risk) and is distributed in 3 possible levels of hierarchizing systems (critical, semi critical and not critical for the organization). Next, the results of the RQCM tool practical application are summarized: 12 equipment evaluated in PRATS Costa RICA (November 2021).

Table 6 Summary of results obtained in the criticality calculation (TCR). (12 systems evaluated)

Equipment	FF	PI	OF	MC	HSE	C	TCR
Satisloh Layoutblocker-PRA blocking machine	3	1	1	1	1	3	9
Laser Micromac 3D UV Laser Rxe 200	1	3	4	1	1	14	14
Satisloh Auto-Flex polishing machine	3	3	1	1	1	5	15
Satisloh VFT-ORBIT generator	2	5	4	1	3	24	48
LOH Lens lacquer machine	2	3	2	1	3	10	20
FISA CS20 4 Plus + R02 ultrasonic washer machine	1	5	4	1	3	24	24
SATTE band Orders Workshop	2	1	2	1	1	4	8
Satisloh Multilayer MC-380-X	2	5	4	1	3	24	48
MEI Bisphera XDD-TBA beveling machine	**3**	**10**	**4**	**1**	**3**	**44**	**132**
SAT Eudepro Tecnic band Beveling and Assembly	2	5	2	1	1	12	24
Enduro Coating SCL CDC 1000PP2	1	5	4	1	3	24	24
Robotic AR Type MCVP8_V2	2	3	2	1	1	8	16

Table 7 Summary of results obtained in the RQCM matrix. (12 systems evaluated)

Department	Equipment	CTR	Ranking
BEVELING AND ASSEMBLY	MEI Bisphera XDD-TBA beveling machine	132	CRITICAL
ORDERS WORKSHOP	Satisloh Multilayer MC-380-X	48	SEMI-CRITICAL
ORDERS WORKSHOP	Satisloh VFT-ORBIT generator	48	SEMI-CRITICAL
ORDERS WORKSHOP	FISA CS20 4 Plus + R02 ultrasonic washer machine	24	NON CRITICAL
BEVELING AND ASSEMBLY	SAT Eudepro Tecnic band Beveling and Assembly	24	NON CRITICAL
TREATMENTS	Enduro Coating SCL CDC 1000PP2	24	NON CRITICAL
ORDERS WORKSHOP	LOH Lens lacquer machine	20	NON CRITICAL
CONTROL	Robotic AR Type MCVP8_V2	16	NON CRITICAL
ORDERS WORKSHOP	Satisloh Auto-Flex polishing machine	15	SEMI-CRITICAL
ORDERS WORKSHOP	Laser Micromac 3D UV Laser Rxe 200	14	NON CRITICAL
ORDERS WORKSHOP	Satisloh Layoutblocker-PRA blocking machine	9	SEMI-CRITICAL
ORDERS WORKSHOP	SATTE band Orders Workshop	8	NON CRITICAL

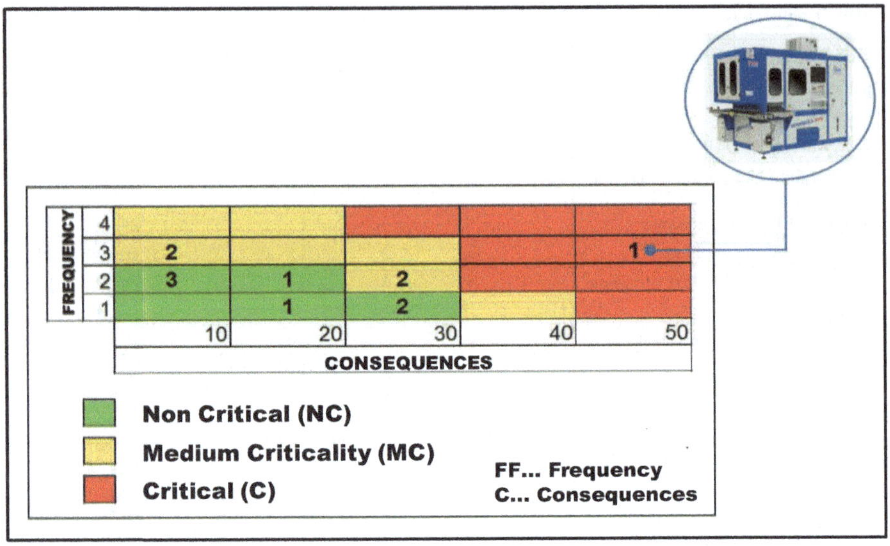

Fig. 7 Summary of all the systems evaluated in the RQCM criticality matrix. Source: Rodríguez, 2021

- 1 item in the critical system area (C) (8,33%)
- 4 equipment in the semi critical area (SC) (33,33%)
- 7 equipment in the non-critical area (NC) (58,33%)

The selection of the RQCM model was justified by the effectiveness and ease of implementation of this technique in the process of the priorization. The RQCM, allows to quickly estimate the factor of frequency and consequence of failures, which can help guide the effective selection of critical equipment. The main limitation of RQCM is associated with the minimum existing information and its low quality (because it is a new production line). It is important to mention that the organization understands and recognizes the technical limitations of the RQCM and its impact on the final results of the case study presented in this report (Rodríguez, 2021). For the application of the RQCM, the organization formed a working group made up of the following people (4 people):

- A leader of the RQCM application (Reliability Engineering).
- Two experts in the types of equipment to be evaluated (Process and Quality Engineering)
- An expert in industry 4.0 (Automation and Control Engineering)

The results of the 12-equipment criticality evaluation are the beginning of an optimization process. After this analysis, in the next phase of the Operational Reliability Optimization Project, the equipment: ***MEI Bisphera XDD-TBA beveling machine,*** that remained in the area of high criticality (C), will be taken; and the following reliability and risk engineering methods will be applied (Rodríguez, 2021; Yoon et al., 2019):

- RCA (Root Cause Analysis)
- RCM (Reliability Centered Maintenance)
- RAM-A (Reliability, Availability and Maintainability-Analysis)
- CRBA (Cost Risk Benefit Analysis)

5 Assessment Recommendations

A key success factor in the implementation of Critical Analysis, is that senior managers who have the authority to make decisions and act on the recommendations of the investigation team, should be involved. An action plan for the implementation of additional risk control measures is the desired outcome of a thorough investigation. The action plan should have SMART objectives: Specific, Measurable, Agreed, and Realistic, with Timescales (Parra et al., 2021b; Viveros-Gunckel et al., 2020). Deciding where to intervene requires a good knowledge of the organization and the way it carries out its work. For the risk control measures proposed to be SMART, management, safety professionals, employees and their representatives should all contribute to a constructive discussion on what should be in the action plan. Not every risk control measure will be implemented, but the ones accorded the highest priority should be implemented immediately. In deciding your priorities, you should be guided by the magnitude of the risk ('risk' is the likelihood and severity of harm). Ask yourself 'What is essential to securing the health and safety of the workforce today?, What cannot be left until another day?, How high is the risk to employees if this risk control measure is not implemented immediately? If the risk is high, you should act immediately. You will, no doubt, be subject to financial constraints, but failing to put in place measures to control serious and imminent risks is totally unacceptable. You must either reduce the risks to an acceptable level or stop the work. For those risks that are not high and immediate, the risk control measures should be put into your action plan in order of priority. Each risk control measure should be assigned a timescale and a person made responsible for its implementation. It is crucial that a specific person, preferably a director, partner or senior manager, is made responsible for ensuring that the action plan as a whole is put into effect. This person does not necessarily have to do the work him or herself but he or she should monitor the progress of the risk control action plan. Progress on the action plan should be regularly reviewed. Any significant departures from the plan should be explained and risk control measure rescheduled, if appropriate. Employees and their representatives should be kept fully informed of the contents of the risk control action plan and progress with its implementation (Parra et al., 2021a).

6 Final Considerations

When carrying out a correct application of qualitative methods of criticality analysis can help both managing and technical levels to make more efficient decisions, directly addressing both economic and human resources in the processes related to the operation and maintenance of industrial assets (Parra & Crespo, 2020a, b; Villar Fidalgo et al., 2018). It is important that maintenance management understand that criticality models to design or use should be aligned with business objectives and not make the mistake of developing criticality tools where only particular maintenance process factors are included. With respect to the latter point, using criticality models based on the Risk factor analysis, it is very interesting, since the Risk Analysis process allows to evaluate the impact of the factors inherent in the maintenance process and to add the assessment of factors such as: production, quality, production losses costs, safety, and environment, among others. Regarding maintenance management, the results of a semi-quantitative criticality analysis process will enable the development of maintenance strategies and management tools with a risk-based optimization approach and its impact on the business.

Below are presented some strengths and weaknesses which are important to consider at the time of the development of a Qualitative Criticality Risk-Based Model (Parra et al., 2021c):

Strengths

- It is a good time to unify the criteria around the risk analysis process (this helps standardize the risk prioritization scenarios of business systems/processes).
- The Qualitative Risk Matrix Model is a simple technique and very easy application (its implantation is fast) requiring only to consult connoisseurs and experts in the systems and the business process to audit.
- It introduces and disseminates the concept of Risk (indicator that allows to integrate the frequency factors and consequences of the failures on safety, the environment, the quality, the operations of the business, etc.).
- It does not require additional (high-cost) resources from the company or business, with the exception of the time spent by the experts to develop the model adjusted to the needs of the organization under study and this be as real as possible.

Weaknesses

- The qualitative methods of Risk generate a high level of uncertainty, therefore we must be very careful with the criteria to evaluate within the model and with the expectations of the management (stakeholders) of the industrial sector where the methodology of criticality analysis is being used, so it is highly recommended for experts to evaluate with the greatest possible objectivity for best results.
- The actual improvements that are obtained from a criticality analysis process, in this case of the qualitative matrix of risk, are going to depend on the subsequent actions that are generated on the critical systems, after having done the criticality analysis, this methodology depends on the subsequent application of other

improvement methods, for example: RCA (Root Cause Analysis), RCM (Reliability Centered Maintenance), CRBA (Cost Risk Benefit Analysis), among others.
- It depends a lot on the information available (key factor to recommend: having expert people and truthful backing information), to such a point, that a risk may be omitted if the starting data are incorrect or incomplete, in addition to overestimating or undervaluing productive systems.
- Being a qualitative, although systematic technique, there is no real assessment of the frequency and consequences of the failures (it is only a qualitative estimate of the reality lived in the business, in this case the lessons learned from the PRATS company to apply them to the new Costa Rican laboratory).

Finally, the results obtained from the effective application of the methodology RQCM (Risk Qualitative Criticality Matrix), guide the Industrial Assets Managers to make decisions more efficiently and with a lesser degree of uncertainty, helping to maximize the profitability of manufactured products in the PRATS Costa Rica factory throughout their entire life cycle.

References

Chiu, Y. C., Cheng, F. T., & Huang, H. C. (2017). Developing a factory-wide intelligent predictive maintenance system based on industry 4.0. *Journal of the Chinese Institute of Engineers, 40*(7), 562–571. https://doi.org/10.1080/02533839.2017.1362357

Crespo, M. A. (2007). *The maintenance management framework. Models and methods for complex systems maintenance*. Springer Verlag. https://doi.org/10.13140/RG.2.2.16765.38884

Crespo Márquez, A., Moreu de León, P., Gómez Fernández, J. F., Parra Márquez, C., & López Campos, M. (2009). The maintenance management framework. *Journal of Quality in Maintenance Engineering, 15*(2), 167–178. https://doi.org/10.1108/13552510910961110

Enap Sipetrol. (2015). Matriz de Evaluación y Gestión de Riesgos, Yacimientos Pampa del Castillo – La Guitarra. ENAP INF-10-2015-CONF1.

Enap Sipetrol. (2016). Proceso de definición de criticidad desarrollado para los activos de producción Off-Shore, Magallanes. ENAP INF-10-2015-CONF2.

González-Prida, V., Parra, C., Gómez, J. F., & Crespo, A. (2012). *Audit to a specific study scenario according to a reference framework for the improvement of the guarantee management*. In G. Berenguer & G. Soares (Eds.), *Advances in safety, reliability and risk management*. https://doi.org/10.13140/RG.2.2.35353.65123

Grupo PRATS. (2021). Sobre nosotros. https://www.grupoprats.com/lentes/espanol/portada_120_1_ap.html

Junietz, P., Bonakdar, F., Klamann, B., & Winner, H. (2018). Criticality metric for the safety validation of automated driving using model predictive trajectory optimization. In *Proceedings of the 21st international conference on Intelligent Transportation Systems (ITSC)* (pp. 60–65).

Li, W., & Wright, M. (2019). Negative emission energy production technologies: A techno-economic and life cycle analyses review. *Energy Technology, 1900871*. https://doi.org/10.1002/ente.201900871

MEI SYSTEM. (2020). Máquinas para el corte de lentes/MEI BISPHERA XDD-TBA. https://meisystem.com/es/

Neurohr, C., Westhofen, L., Butz, M., Bollmann, H., Eberle, U., & Galbas, R. (2021). Criticality analysis for the verification and validation of automated vehicles. *IEEE Access, 9*, 18016–18041. https://doi.org/10.1109/ACCESS.2021.3053159

Parra, C., y Crespo, A. (2015). *Ingeniería de Mantenimiento y Fiabilidad Aplicada en la Gestión de Activos. Modelo de Gestión del Mantenimiento (MGM)*. 2da, ed. INGEMAN, Escuela Superior de Ingenieros Industriales. https://doi.org/10.13140/RG.2.2.29363.66083

Parra, C. y Crespo, M. (2018). Nota técnica 5: Métodos de Análisis de Criticidad y Jerarquización de Activos. INGEMAN, ETSI. https://doi.org/10.13140/RG.2.2.21197.87524

Parra, C., y Crespo, A. (2019). Nota técnica 4: Técnicas de Auditoría aplicadas en los procesos de Gestión del Mantenimiento. INGEMAN. https://doi.org/10.13140/RG.2.2.10169.60003

Parra, C., y Crespo, A. (2020a). Nota técnica 1: Introducción a un modelo integral de Gestión del Mantenimiento (MGM). INGEMAN, ETSI. https://doi.org/10.13140/RG.2.2.13046.63049

Parra, C., & Crespo, A. (2020b). *Introducción al Modelo Integral de Gestión del Mantenimiento y de la Confiabilidad alineado con el enfoque de la norma: UNE 16646 (Mantenimiento en la Gestión de Activos)*. INGEMAN – Universidad de Sevilla. Artículo No.3, pp. 5–10.

Parra, C., Viveros, P., Kristjanpoller, F., Crespo, A., & González-Prida, V. (2020a). *Modelos de auditoría para los procesos de gestión de activos, mantenimiento y confiabilidad*. Sector de Transmisión de Electricidad. INGEMAN. https://doi.org/10.13140/RG.2.2.32132.14721/1

Parra, C., González-Prida, V., Candón, E., De la Fuente, A., Martínez-Galán, P., & Crespo, A. (2020b). Integration of asset management standard ISO55000 with a maintenance management model. In M. A. Crespo, D. Komljenovic, & J. Amadi-Echendu (Eds.), *14th WCEAM proceedings. WCEAM 2019* (Lecture notes in mechanical engineering). Springer. https://doi.org/10.1007/978-3-030-64228-0_17

Parra, C., Tino, G., Parra, J., Crespo, A., Viveros, P., Kristjanpoller, F., & González-Prida, V. (2021a). *Metodología básica de análisis de riesgo para evaluar la criticidad de activos industriales*. Línea de Manufactura de Envases Biodegradables. Edita INGEMAN, ETSI. https://doi.org/10.13140/RG.2.2.10422.52802/2

Parra, C., Viveros, P., Kristjanpoller, F., Crespo, A., González-Prida, V., & Gómez, J. (2021b). *Técnicas de auditorías para los procesos de: mantenimiento, fiabilidad operacional y gestión de activos (AMORMS & AMS-ISO 55001)*. INGEMAN. https://doi.org/10.13140/RG.2.2.35842.61124/4

Parra, C., Crespo, A., Parra, J., Viveros, P., Kristjanpoller, F., González-Prida, V., & Balda, A. (2021c). *Técnica de Jerarquización de Activos MCCR: Matriz de Criticidad Cualitativa de Riesgo*. Unidad de Craqueo Catalítico. INGEMAN, ETSI. https://doi.org/10.13140/RG.2.2.31889.15209/1

Rodríguez, J. (2021). *Propuesta de diseño para un modelo de gestión de mantenimiento con acercamiento a técnicas de la Industria 4.0 para la organización: PRATS C/Industrias de Óptica S.A.* (pp. 44–84). Instituto Tecnológico de Costa Rica (ITCR/TEC).

SATISLOH. (2020). The art of making lenses. Recuperado de: https://www.satisloh.com/

Villar Fidalgo, L., Crespo Márquez, A., González Prida, V., De La Fuente, A., Martínez-Galán, P., & Guillén, A. (2018). Cyber physical systems implementation for asset management improvement: A framework for the transition. In *Safety and reliability–safe societies in a changing world* (pp. 3063–3069. ISBN: 9781351174664). CRC Press. https://doi.org/10.1201/9781351174664

Viveros-Gunckel, P., Kristjanpoller-Rodríguez, F., Parra-Marquez, C., Crespo-Márquez, A., & González-Prida-Díaz, V. (2020). Audit models for asset management, maintenance and reliability processes. Case study: Electricity transmission sector. *DYNA Management, 8*(1), 14. https://doi.org/10.6036/MN9826

Yoon, J. T., Yooung, B., Yoo, M., Kim, Y., & Kim, S. (2019). Life-cycle maintenance cost analysis framework considering time-dependent false and missed alarms for fault diagnosis. *Reliability Engineering & System Safety, 184*, 181–192. https://doi.org/10.1016/j.ress.2018.06.006

Carlos Parra is currently Professor at the Mechanical Department, UTFSM Chile. His first 18 years of experience were directly in the maintenance area in the oil, gas, manufacturing, and petrochemical sectors. He completed a master's degree in maintenance engineering (University of the Andes, Venezuela) and a Ph.D. in Industrial Engineering (University of Seville, Spain). In the last 8 years, Prof. Parra gained experience in the manufacturing, food and beverage, mining, energy and oil sectors, participating in maintenance and operations optimization projects in the main companies throughout America. Total of publications on Research-Gate: 253 (October 2023). Books published: Cases on Optimizing the Asset Management Process, 2022, IGI Global, United States and "Ingeniería de Mantenimiento y Fiabilidad aplicada en la Gestión de Activos", 2015, INGEMAN, Spain.

Juan Rodríguez graduated from the Costa Rica Institute of Technology with a degree in Industrial Maintenance Engineering, earning honors for the thesis developed in the field of ophthalmic lens manufacturing (Industrias PRATS Costa Rica). Since 2021, he has been working as a maintenance engineer in Costa Rica's most significant industrial sectors, applying operational excellence tools and reliability engineering, such as Lean Maintenance, RCM, RAM, and TPM.

Adolfo Crespo Márquez is currently Full Professor at the Department of Industrial Management, School of Engineering of the University of Seville. He holds a PhD with Honors in Industrial Engineering from this same university. He is included in the Stanford University's list of world top-scientists in 2023 and he received the Spanish Maintenance Award from the Spanish Maintenance Association (AEM) in 2020. With research works published in many top journals, he is the author of 8 books and editor of 6 books with Springer, Aenor, IGI Global and INGEMAN, about maintenance, warranty, supply chain and assets management. Prof. Crespo is Fellow Director of ISEAM, Editor-in-Chief of EAMR Springer Book Series and led the Spanish Research Network on Assets Management and the Spanish Committee for Maintenance Standardization (1995–2003). He also leads the SIM research group related to smart maintenance systems and has extensively participated in many engineering and consulting projects for different companies, for the Spanish Departments of Defense, Science and Education as well as for the European Commission (IPTS). He is the President of INGEMAN (a National Association for the Development of Maintenance Engineering in Spain) since 2002.

Vicente González-Prida obtained his PhD in Industrial Engineering from the University of Seville, specializing in Asset Management and Maintenance. With 20 years of global industry and academic experience, he worked as an international project manager, focusing on new technologies for Customer Service and ILS. Currently a lecturer and researcher at the University of Seville, Spain, He has authored and edited books on Maintenance Modelling, RAMS, Asset Management, as well as After Sales Management. Actively participating in conferences like WCEAM and ESREL, he publishes in high impact journals (RESS, PP&C, etc.) and his academic work has been recognized with the Prize of Doctorate from University of Seville, National Awards for PhD Thesis among others. He is a member of the Spanish Institute of Engineering and other institutions.

Pablo Viveros holds a Ph.D. in Mechanical and Industrial Engineering from the University of Seville, Spain. Holds a degree in Industrial Civil Engineering and a Master's in Asset Management from USM. Currently serving as an Academic in the Department of Industries at UTFSM, actively contributing to WoS scientific research and engineering projects for both national and international industries. Specializes in consulting for project development and leveraging technological tools for asset management and maintenance.

Fredy Kristjanpoller is currently a Professor at the Industrial Engineering Department at Universidad Técnica Federico Santa María, Chile. He received his Ph.D. in Mechanical Engineering and Industrial Organization at the University of Seville, Spain. His research and teaching interests are in operations, asset management, reliability, maintenance strategies and risk analysis. He has been published in several journals, including Reliability Engineering and System Safety, Complexity, Renewable Energy, Journal of Risk and Reliability and others.

Jorge Parra is a Mechanical Engineer, graduated with honors as a Sigma Lambda member in Mechanical Engineering with a specialization in Energy and Environment from the Technological University of Panama (2017–2022), receiving recognition and a medal of honor for outstanding academic performance. Currently working as a Support Engineer at Ingecon S.A since March 2022, providing virtual support for development and optimization projects of maintenance plans, applying reliability techniques, and producing various technical reports. Previous experience as a Project Assistant at Moldeados Panameños S.A (MOLPASA), where he conducted an energy audit, thermography, and life cycle analysis. Also participated in the Virtual Scientific Initiation Program at UTP with a project on the use of eggshells and wastewater sludge as organic fertilizer in tomato production.

Factors Affecting the Quality of Network Services in Emerging Telecoms Operating Environment and Markets

Charles Okeyia and Nuno Marques Almeida

Abstract As an emerging market, the telecoms sector in Nigeria has undergone a considerable increase in teledensity, internet usage and consumer base over a decade and is still on exponential growth. However, the consequence of this increase in growth has been a continuous degradation of telecom network quality of service (QoS), which has impacted subscribers' customers' needs, satisfaction, expectations and added value services. In exploring the quality of services (QoS) issues, the asset performance is not meeting the agreed key performance indicators (KPIs) on power availability (PA), a critical KPI which is affected by asset maintenance activities. Therefore, this paper focuses on the technical and human factors of asset management and maintenance practices. The methodology used in this paper is the quantitative and qualitative approaches with a systematic review of related literature on the research context. The primary data sources are through a structured survey questionnaire and semi-structured interviews. The secondary data source is the systematic literature review on related journal articles to the research subject matter. The paper used the statistical package for the social sciences software (SPSS 29) and Nvivo software for the data analysis. The research results and findings indicate critical maintenance strategic differences in existing asset maintenance activities and operations, cost pressure, and complex operating environments and markets that could be explained through intelligent and digitalised asset management and maintenance strategies. The systematic review results indicate the advancement of asset maintenance strategies to support maintenance planning, asset real-time monitoring and management, as the existing maintenance practice did not match the intelligent-based approach drawn from the concept of Industry 4.0R.

C. Okeyia · N. M. Almeida (✉)
Instituto Superior Tecnico (IST), Lisbon, Portugal
e-mail: charles.okeyia@tecnico.ulisboa.pt; nunomarquesalmeida@tecnico.ulisboa.pt

A. Crespo Márquez et al. (eds.), *Advances in Asset Management: Strategies, Technologies, and Industry Applications*, Engineering Asset Management Review 3, https://doi.org/10.1007/978-3-031-52391-5_2

1 Introduction

Increasing teledensity and customer satisfaction in emerging telecom markets requires intelligent and digitalised asset management and processes. This includes more flexible and cost-effective operations management strategies due to the cost pressure on operations and maintenance and the effect of human and environmental challenges on asset management activities. Public electricity (power grid) is a crucial issue in Nigeria's operating environment that impacts telco infrastructure asset management regarding reliability, performance and operation cost. Over seventy-five per cent (75%) of the telecom infrastructure and asset base stations are off public electricity. These infrastructure and assets operate mainly on alternating current diesel generators (ACDG) and direct-current diesel generators (DCDG), with high operating costs and the integration of green or clean energy solutions like solar and hybrid batteries, which also are prone to functionality issues because of theft on the components such as panels and batteries.

In addition, the asset operations and maintenance strategies need to meet the requirements of the likely asset lifecycle that addresses the infrastructure and asset challenges based on the existing operations and maintenance conditions. Thus, as key maintenance enablers, operations and maintenance needs, intelligent and digitalised maintenance strategies must have improved flexibility and efficiency in ensuring network quality of service and asset performance that addresses the network quality of service (see Fig. 1).

However, the existing asset management and maintenance practices are reactive, time-consuming, and not responsive or intelligent enough to address network quality of service challenges, thereby creating issues with network availability, which is a critical KPI for network quality of service. The present asset management and maintenance activities rely greatly on human interventions, individual skills and unintelligent escalation processes that increase mean-time-to-repair and operating expenditure (OPEX) based on poor work attitude and non-compliance to Predictive-based maintenance strategies. These factors, such as intermittent asset outages, field technicians' frequent visits to sites, and poor maintenance practices, are key drivers of asset performance, which is a critical factor affecting the quality of service.

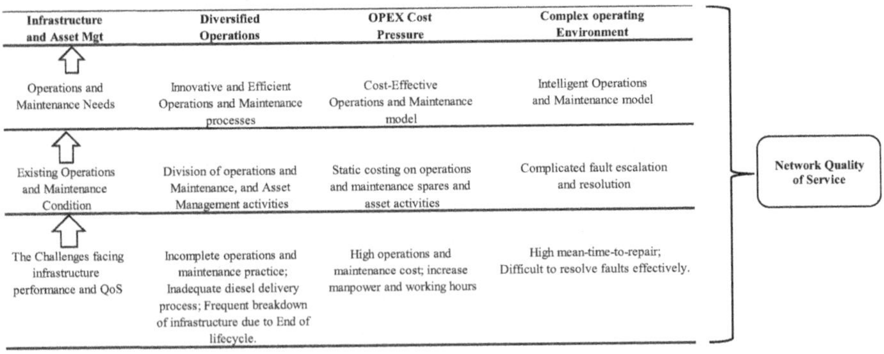

Fig. 1 Flowchart of key maintenance enablers

In the telecoms domain, quality of service (QoS) is the mechanism in the network systems that control traffic and ensure critical application performance and availability with limited network capacity or a set of parameters connected with traffic performance in telecoms (Opara et al., 2021). It is measured based on network availability, congestion or packet loss. On the other hand, quality of service aims at the technical characteristics of a provider's perspective (Raj & Basar, 2019) and service received by the customer.

The network quality of service indicates that telecom infrastructure and asset performance on quality of service is multidimensional with various reported dimensions (Olde Keizer et al., 2018), and factors such as intermittent asset outages and poor compliance to routine planned preventive maintenance activities are challenges affecting asset performance that in turn impact QoS. The level of automation is mainly low, such as the integration between various asset redundancy procedures, such as synchronisation of the public grids and the diesel generators 1 or 2. Thus, the performance of these infrastructures and assets impacts the agreed key performance indicators, such as network availability, mean-time-to-repair and quality of service.

2 Literature Review

Despite several pieces of extant literature that support the importance of asset management, almost no published research addresses infrastructure and asset management and maintenance activities that impact QoS in Nigeria's telecoms context. Given the importance of sustaining and addressing the issues with the network quality of service in the research domain, our purpose is to explore the factors affecting network quality of service in Nigeria as an emerging telecom operating environment and market. This understanding explains how the quality of network services greatly influences customers' perceptions, needs and expectations. It is, therefore, required to identify the factors affecting the QoS. In achieving this exploration, the paper focused on the technical, human and operating environmental areas that have not been researched in this context, such as infrastructure and asset management and maintenance strategies. Previous studies (Lewis & Booms, 1983; Grönroos, 1998; Parasuraman et al., 2015; Sugeng, 2016; Kotler et al., 2019) have overlooked the impact of asset impact on quality of service but rather focused on quality service as related to customer satisfaction, marketing, sales, promotions, and regulatory and government interventions.

Customer satisfaction is typically ascribed to issues with network availability and quality of service, thus impacting customer's needs, expectations, value-added services and dissatisfaction. However, much of the existing work on similar emerging operating environments and markets focused on applied technologies for powering assets, such as hybrid solar, batteries, and other green systems (Oviroh and Jen (2018). Other research works aim at consumer dissatisfaction (Opata, 2013a, b), connectivity and the digital ecosystem (Adame, 2021). However, various research works agreed that no contemporary telecoms markets could be developed and continued short of an efficient telecoms infrastructure and service (ITU, 2019; Vu et al., 2020). These existing research works did not focus on infrastructure and asset management or maintenance roles in sustaining network values such as QoS and network availability. This understanding justified the broad divergence in the description of network quality from a general business perspective as services without defects (Brady & Cronin, 2001) and the concepts that included internal resources and service (Duggal & Verna, 2013). However, from a customer's critical perspective, Parasuraman et al. (1988) described quality as a global attitude or judgement relating to the superiority of the service.

In contrast, Lewis and Booms (1983) defined quality of service as customers' expectations of the performance attained from the services offered. These descriptions could be linked to the paper's concerns about subscribers' needs and expectations of the added value services. However, the performance of service could be quantified and examined based on the tangibility of the activities from the infrastructure and asset management and maintenance practices that provide the service. From the physical environment and infrastructure and asset perspective, Almossawi (2012) noted that internal company policies, service challenges, customer satisfaction and organisation position are attributes that affect QoS. In the context of this paper, we focus on the infrastructure and asset management and maintenance concept that will develop intelligent and digital systems driven by predictive-based maintenance and condition-based maintenance strategies to enhance asset performance in an effort to provide stable quality network service.

Notwithstanding the applied conventional maintenance strategy of planned preventive maintenance cycles in the research context based on the recommended manufacturer's manual book followed by the operators, these infrastructures and assets break down intermittently, causing poor network availability and stability of network quality of services, increasing operating expenditure which impacts asset performance. Nakajima (1988) maintained that asset performance is the measurement and identification of outages of critical business characteristics such as availability, performance and quality. This explains the understanding that network availability is expressed as the uptime and downtime of infrastructure and assets (Kehinde et al., 2017). A better experience and performance of network availability influence QoS (De Azevedo, 2019) as this perspective relates to asset management and maintenance strategies.

2.1 Telecoms Industry

Telecoms engineering is the field of business dealing with ICT systems that provide network services and other internet access. Network quality of service in the context of this paper has drawn attention because of the impact on network providers' performance. Quality of service is generally a determinant of how satisfactory the extent of delivered service meets the customer's expectations and needs (Santos, 2003). This description is outlined based on perceived service quality by the users (Grönroos, 1984). Also, this description was supported by Parasuraman et al. (1988) as a general assessment of a particular service organisation that results from contrasting that organisation's performance with users' overall expectation of the organisation's performance indicators. Unlike assumed better quality of services, which can be assessed with some objectivity, QoS is measured based on certain defined key performance indicators (KPIs) that are majorly dependent on network power availability (PA). Power availability is an outcome of the passive asset performance that supports the active performance under which QoS is quantified and measured. However, Ghobadian et al. (1994) argued that quality of service has characteristics such as inseparability of service and consumption, intangibility, heterogeneity and decreased capability measures quality of service as a very multifaceted issue. Thus, because of the lack of objective measures, organisations rely on users' views of service quality to determine their strengths and weaknesses and put appropriate strategies. This conclusion is not the case in this paper; rather, the paper focuses on quantifiable factors that affect the quality of service.

Given these insights on quality of service and the focus of this paper on infrastructure and asset management and maintenance activities as factors that affect QoS, we classify active and passive asset components and elements. The passive assets ensure high power availability and reliability that supports the active components related to network quality of service. Availability in this context is the capacity to fulfil and access the network (Hedvall & Paltschik, 1991; Kehinde et al., 2017), and it is determined by the level and extent to which the network services are everywhere. On the other hand, availability is the complement of reliability; thus, they are the core factors of quality of service, considering the key technical characteristics in this domain. This understanding explains that reliability is the probability that the services will be actively operating the exact function in an absolute setting for all duration. For instance, the expected asset lifecycle or the asset means-time-to-repair (MTTR) between outages can originate from this likelihood duration. This is because asset failures, measured at a variation of times, prevent customers' accessibility to the network. A particular challenge in this context is to increase and sustain network availability through passive asset performance, which necessitates new maintenance approaches for better functionality of the asset.

2.2 Network Service Quality Parameters

Network availability in the context of the telecom is the parameter that defines the quality of service (QoS) from the network and service provider perspective. In contrast, the basic requirement from the user's perspective is based on the accessibility of the network. Nurysh et al. (2019) agree with Abd-Elrahman (2019) on the findings that perceived value and service quality have a positive relationship with user satisfaction, where service reliability is a factor that affects the quality of service. The perceived network service quality is the attitude and behaviour concerning the dominance and advantage of the service compared to the context (Parasuraman et al., 1988, 2015). This understanding involves experiences related to customers' beliefs about the asset or utility arising from services which they experienced

(Parasuraman et al., 1988, 2015). Abd-Elrahman (2019) concludes that the role of perceived quality of service is determined by the settings under various factors that affect QoS evaluation and user satisfaction. Garin (1987) and Abd-Elrahman (2019) support this proposition with identified critical characteristics that a service must have to be considered of high quality, such as performance, reliability, durability, perceived and features.

Conversely, we can conclude from the literature that quality of service (QoS) is dependent on various aspects of the interconnect between the passive assets and active assets, which the network providers manage. Thus, this paper focused on the passive asset that supports the active assets in providing the necessary network availability, thus the factors from asset performance affecting network quality of services. For critical assets such as generators, power systems, solar and green solutions, air-conditioning units and hybrid systems, the maintenance procedure has to face different typical faults management and different operations. Thus, in this context, all the assets are combined inside each base station. The maintenance process is specifically directed at maintaining the infrastructure and assets involved in the activities. Typical infrastructure and asset outages are classified as generator faults such as mechanical, electrical, injector or fuel pump failure, high temperature or low water, and high and low frequencies. Automation systems failure, such as synchronisation failure, alarms failure malfunction. Fuel supply failures such as low diesel levels, adulterate quantity and quality, and theft. Cooling faults such as air conditioning unit faults.

Additionally, telecom infrastructure and asset management and maintenance activities are characterised by routine and non-routine tasks, which involve the need for rational maintenance strategies for planning and monitoring. The cost of routine maintenance in this context is high because of the 24-h running and requires logistics, spare replacement and necessary capacities.

2.3 Maintenance Practice

In this researched environment, the supply and provision of the public grid or electricity are unreliable compared to other developing nations where telecom services infrastructure and assets are operating on the public grid and electricity. Excluding the regular load shedding of public electricity, the power availability of public electricity in this context supplies to telecom infrastructure and assets is less than 10% of the duration, thereby increasing the paper context infrastructure and assets dependence on diesel generators and other solutions such as hybrid and solar systems. This dependence on diesel-generating sets has created a huge OPEX cost for telco maintenance organisations (Danbatta & Zangina, 2022). In addition to the poor availability of electricity in this telecom infrastructure and assets, telco maintenance organisations are also faced with other operational challenges and issues. For instance, diesel pilferage has impacted operational costs (OPEX) and theft of passive equipment. However, the plan towards reducing the diesel intake has been inconsistent with other operations and maintenance associates' interests, thus impeding the effective execution of the other power options. Given this maintenance context, telco maintenance organisations must strengthen their operations and maintenance systems and processes to reduce the impact on infrastructure and asset performance and improve the quality of network services that address customers' needs and expectations.

On the other hand, the importance of this Predictive-based maintenance approach is for optimal planned inspection (Aremu et al., 2018; Marquez et al., 2020; Pais de Almeida et al., 2021). In this manner, prolonged or early interventions, avoidable infrastructure and asset interruption and high MTTR are decreased, and sudden downtime is detected and prevented. Therefore, if the infrastructure asset elements can be monitored constantly through intelligent or digital devices, intervening for maintenance and inspection could be likely only at discrete periods that distinguish between periodic and aperiodic decision instants (Olde Keizer et al., 2018). The Predictive-based maintenance approach involves several tasks such as alarm installation, data and information classification, and asset management activities – high-temperature faults, cooling alarms, power alarms, low diesel levels processing and decision-making. In reference to the intelligent predictive-based maintenance approach, the simulation instruments and intelligence analysis are drawn from artificial intelligence (AI) systems. The AI simulates human intelligence processes through human and machine learning collaboration termed hybrid intelligence (Kamar, 2016; Dellermann et al., 2019) to address human intervention and inappropriate maintenance activities, whereby infrastructure and asset functionality are monitored in real-time. This action is achieved through intelligent or digital like sensors, meters, and vision systems that monitor real-time system performance and collect data to manage uncertainties in infrastructure and asset activities—automating data and real-time escalation of predictive outages saves human time, cost and better decision-making.

2.4 Maintenance Strategies

Maintenance is a required feature of the asset management process in the telecom domain. Maintenance includes technical and human activities related to planning, inspections, condition-based monitoring, routine maintenance, repairs, main and minor overhauling, spare replacement and supervision (Olde Keizer et al., 2018). Although several systematic reviews cover the conventionally used techniques for asset maintenance in the context of this paper in such a comprehensive manner, not much review covered the industrial 4.0R concepts such as simulation, big data, intelligence, digital twin and predictive-based maintenance strategies, except in other contexts. A comparison and summary of the systematic review was based on criteria such as the maintenance strategy that indicates the effect of the presented content on management approaches. For instance, maintenance strategy formation or likely change of existing processes or practices.

A maintenance strategy is categorised based on the duration when a repair on an outage is conducted in relation to the incidence of the breakdown or outage, such as preventive, corrective and predictive maintenance strategies (Molęda et al., 2023). However, based on these understandings, the existing maintenance strategies (preventive and corrective) in this paper context have not addressed the issues of asset performance, which in turn impact the quality of service due to certain maintenance culture, implementation and environmental challenges. Thereby provocating concerns on how contemporary technologies support existing asset maintenance strategies and practices. Inappropriate maintenance strategies and practices reduce and impact performance (Satish & Anil, 2017); likewise, intelligent and Predictive-based maintenance strategies increase asset performance and availability and reduce maintenance costs or operating expenditures (Bradbury et al., 2018). thus, the increase in the application of intelligent or predictive-based maintenance strategies.

Additionally, the corrective maintenance strategy suggests conducting activity after a breakdown has happened. This method reduces the asset maintenance or servicing cost but with the risk of intermittent asset failures because of the change in the maintenance interval, and should be more appropriate for non-critical, simple, repairable asset faults. Intermittent asset failures create a loss of revenue, increased operating expenditure (OPEX), and increased mean-time-to-repair (MTTR) as a result of unplanned breakdown and downtime. All these are factors that impact asset performance, which in turn impact the network quality of services.

In contrast, the predictive-based maintenance strategy is performed when it is needed, normally abruptly prior to anticipated breakdown. The significance of the predictive-based maintenance practice is to envisage and predict the condition of an asset based on recurring analysis or identified features. This understanding suggests that predictive-based maintenance is a kind of condition-based maintenance that envisages and foresees future performance based on present and historical signs or pieces of evidence (Olde Keizer et al., 2018; Molęda et al., 2023). The use of this strategy leads to a decrease in unplanned and planned breakdowns.

2.5 *Computational Devices for Maintenance Planning*

Several approaches have been established to optimise the maintenance strategies for asset management support in planning and operating cost reduction (Rinaldi et al., 2016). This tool comprises inputs, simulations, mechanisms and outputs that simplify the maintenance planning and approaches, which was initially developed for the energy domain that allows flexibility across different technologies, specifically wind turbines and energy converters. The critical assets in this paper context are energy and power-driven based on the challenges of providing power for the equipment, making the power availability a key KPI. However, an adjustment is required to make these tools appropriate to capture and simulate the dynamic process of asset maintenance activities over time (Rinaldi et al., 2016), usually the asset lifecycle and explore various aspects of techniques to detect the difficulty and ineffectiveness of the maintenance strategy and propose likely areas for enhancement. Shafiee (2015a, b) noted that each instrument or approach is individually created to describe one or several characteristics of the asset activities and to span various planning perspectives.

Most of these approaches and tools simulate the dynamic process of asset maintenance over time, usually the asset lifecycle and explore various techniques to detect the difficulty and ineffectiveness of the maintenance strategy and propose likely areas for enhancement. Shafiee (2015a, b) noted that each instrument or approach is created to describe one or several characteristics of the infrastructure and asset activities and span various planning perspectives. Although some tools include optimisation features, other means are restricted to characterisation, usually assessing the agreed key performance indicators (KPIs). The mechanisms described the guidelines for the simulation of the maintenance activities. The inputs and constraints include the task descriptions for the individual method and the whole infrastructure assets maintenance (Rinaldi et al., 2016). The outputs offer a technical and cost-effective task evaluation during the simulated cycle. However, several of these maintenance models are adaptable and can certainly be tailored toward intelligent or digital asset maintenance practices. Figure 2 visually describes the mechanism, inputs, constraints and outputs traditionally considered in a telecoms maintenance simulation tool.

The inputs involve the asset management and maintenance activities that are supposed to sustain the asset performance in order to achieve the KPIs stated as the outputs – power availability of 99.99%, reduced mean-time-to-repair 30 min and reduced diesel outages and diesel line blockages to 0%. If these activities are sustained assets, performance impact on quality of service issues will be addressed.

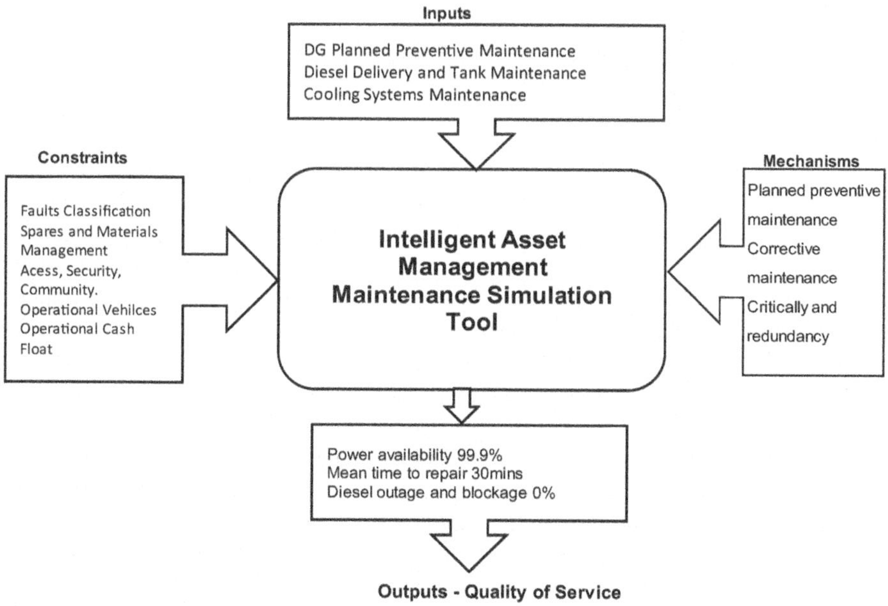

Fig. 2 Maintenance tool. (Rinaldi et al., 2016)

However, with human-centric interventions, certain constraints have been observed to affect asset maintenance activities. For instance, fault escalations, inadequate spares and materials holding such as batteries, air filters or elements, fan belts, erratic load imbalance, low generator frequencies, diesel line blockage and low level, alternator failure, etc., are not readily available in most situations. Additionally, access and security to sites due to landlord and community demand contribute minimal impact. Operational vehicles and float are challenges faced internally by the field operational teams due to issues with individual managed service organisations. These constraints are not significantly a key concern as they are micro-managed at the operational level.

Given the description of the inputs, constraints and outputs, the mechanism covers the maintenance strategies that aim at each asset element independently because of the assumption that downtime, faults, and degradation situations are independent. The outputs offer a technical and cost-effective task evaluation during the simulated cycle. The detailed assessments of the maintenance simulation tools are described as these are examined and categorised based on the objective, attributes, working standard and fundamental methodology. However, several of these maintenance models are adaptable and can certainly be tailored to other asset maintenance practices. A significant aspect of the application and exploitation of intelligent asset management for maintenance planning and implementation is the real-time faults report, intelligent or digital resolution and monitoring of maintenance activities. However, this method is complicated in telecom infrastructure and asset operations because of the comparative uniqueness of the inadequate knowledge of the mechanism.

2.6 Condition-Based and Predictive-Based Maintenance Approach

The condition-based or Predictive-based maintenance approach integrates data-driven reliability simulations with data gathered from the alarm mechanism and condition monitoring structure to create an enhanced maintenance strategy (Hameed et al., 2010). For example, proactive and predictive maintenance could be planned to use the data produced by the alarm mechanism and the knowledge accumulated from the historical data. This perspective is related to Zhao et al. (2019) condition-based maintenance (CBM) approach and Pais de Almeida et al. (2021) optimising the life cycle of physical assets, which typically results in greater availability and reduces maintenance costs since it aims to avoid unplanned outages and prevent avoidable preventive maintenance activities for infrastructure and asset. However, the advantage of CBM remains uncertain in multi-infrastructure systems such as the telecom setting, where opportunistic maintenance strategies can be used. In the current network quality of service context, opportunistic maintenance could be an extra in network management because it aims to classify maintenance activities of several elements to lower maintenance costs (Farinha, 2020), as in support that CBM could be cost-effective (Zhao et al., 2019).

This suggestion happens when necessary, for example, when the infrastructure and asset are still in working condition, in progress, or very late when the infrastructure and asset have broken down, thereby triggering a high mean time to repair (MTTR). In contrast, monitoring diesel consumption and delivery, which contributes to the higher impact, may be problematic for these approaches; thus, AI and human-centric collaboration can be applied through intelligent or digital and systematised maintenance practice. The critical stages of the condition and Predictive-based maintenance approach are:

- Information about the processed signal is obtained from the appropriate data – such as fault identification and analysis.
- Reliability modelling to capture infrastructure asset deterioration, then send a signal to forecast failures. This action comprises fault prediction and subsequent proof with the current database.
- Decisions toward maintenance optimisation involve inspection and planned preventive maintenance.

The Predictive-based maintenance approach is outlined in a flowchart in Fig. 3. It depends on the mixed utilisation of data gathered from alarm devices and intelligent or digital equipment with calibrated reliability models to assist the planned preventive maintenance.

This Predictive-based maintenance approach aims to obtain well-defined, timely and specific signals regarding when infrastructure asset maintenance is essential (Van Horenbeek et al., 2010). On the other hand, the emphasis of this Predictive-based maintenance approach could be on the optimal planned inspection. In this manner, prolonged or early interventions, avoidable infrastructure and asset interruption, and high MTTR are decreased and sudden downtime is prevented. Therefore, if the infrastructure asset elements can be monitored constantly through

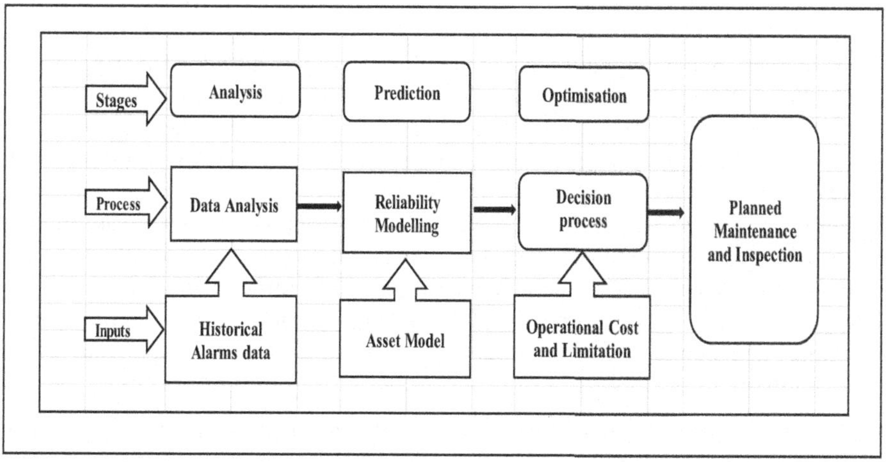

Fig. 3 Flow diagram of predictive-based maintenance approach

intelligent or digital devices, intervening for maintenance and inspection could be likely only at discrete periods that distinguish between periodic and aperiodic decision instants (Olde Keizer et al., 2018). The drawback is that this Predictive-based maintenance approach involves a detailed knowledge and interpretation of multi-system changing aspects, particularly interdependent infrastructure and asset downtime or faults, environmental situations, and associated non-technical consequences (Lawrence & O'Connor, 1995). Moreover, this Predictive-based maintenance approach involves several tasks such as alarm installation, data and information classification, and asset management activities – high-temperature faults, cooling alarms, power alarms, low diesel levels processing and decision-making.

The Predictive-based maintenance approach is critical in the telecoms infrastructure and asset maintenance process because of the various factors and activities – erratic load imbalance, low generator frequencies, diesel line blockage and low level, alternator failure, and operation cost. However, this Predictive-based maintenance approach aims at each infrastructure and asset element independently because of the assumption that downtime faults and degradation situations are independent. Although this proposition could be a generalisation that may perhaps lead to the inaccurate assessment of the infrastructure and asset downtime or MTTR. (Peng et al., 2014; Camci, 2009) posits that fault dependencies between infrastructure and assets could be distinct and outlined as follows;

- Functional or structural dependence – infrastructure and asset breakdown or not functioning as a result of another dependent infrastructure and asset not working, which could be because of technical – failure or maintenance – performance; in such a situation, the maintenance of the non-dependent infrastructure and asset may require pre or post-intervention on the dependent infrastructure and asset.
- Stochastic dependence – the breakdown of one infrastructure and asset impacts another infrastructure and asset. The shared mode in which this could occur is a direct fault–induced damage–failure, owing to a rise in the workload shared until that moment of load sharing and – mutual decline – standard mode. For example, the increase in load or equipment could impact the generator load sharing among various collocated infrastructures and assets in the telecoms base station. In this situation, (Rasmekomen & Parlikad, 2016) noted that the degradation of dependent infrastructure and assets could be modelled using a standard probability distribution.
- Resource dependence – an inadequate quantity of particular resources, such as spare parts, tools and technicians, hampers maintenance.
- Economic dependence – maintenance of multiple infrastructure and assets is either cheaper – positive economic dependence or more costly – negative economic dependence than maintaining the same infrastructure and assets individually.

Equally, the dependencies lead to secondary effects, such as maintenance cost or service and asset lifecycle reduction, because of the various multi-assets that are disassembled elements in the base station. For instance, a power system outage affects the cooling units, security lights and real-time escalation of active components. Therefore, for this reason, (Wang, 2002) posits that proper opportunistic maintenance may not be recurrent or time-based, as the inspection process could support the planned interventions that reveal the condition of the infrastructure and asset. However, this action requires the consideration of two activities. Firstly, the inspections could be fallible, such as failure in detecting faults.

In contrast, the inspections have to be adequately planned, following standardised criteria for maintenance optimisation, such as reliability maximisation and cost minimisation. However, Sheu et al. (2015) noted that inspections could be intermittent in continuous monitoring, both routine or non-routine, and the decision at the correct period may differ due to the asset lifecycle status. Therefore, planned inspection optimisation is a significant characteristic of the complete maintenance planning and decision-making model that can improve infrastructure and asset performance.

2.7 *Diagnosis and Prognosis*

Diagnosis and prognosis are the methods employed to chart the condition-based maintenance data and forecast the future situations and status of the assets (Kang et al., 2019). When the information on the procedure that led the asset to fail is presented, the data will be utilised to identify the failed infrastructure and assets. Diagnosis is applied to analyse the distinctions between normal working conditions and failed conditions. At the same time, the prognosis is employed to foresee when the infrastructure and assets are imminent to break down. These two methods applied numerical or digital models to describe the correlation between fault causes and fault mechanisms (Mathur et al., 2001; Liu et al., 2018; Olde Keizer et al., 2018), as they assist in assessing the past, present (diagnosis) and future (prognosis) conditions of the infrastructure and asset system.

The two critical methods known for diagnosis purposes are the statistical analysis that identifies a signal that can signify unusual infrastructure and asset conditions and exploits computational procedures capable of self-adapting and improving after an early training phase. Rinaldi et al. (2021) noted that genetic algorithms (GAs), Machine Learning (ML) and artificial neural networks (ANNs) are the most employed computational approaches. However, the critical aim of prognosis is to decide the effective residual lifecycle before the infrastructure and asset breakdown happens. This is due to uncertainty because of the stochastic breakdown behaviour of the infrastructure and assets. However, this inference permits an evaluation of the breakdown possibility before an inspection or a planned preventive intervention, which supports optimising the maintenance period. Prognosis could be usually attained by a scientific or physical model of the infrastructure and asset – model-based or physics, by design recognition in formerly obtained data – data-driven through earlier knowledge with the maintenance of the same infrastructure and asset – experience-based, or a combination of hybrid. This option of a prognosis method over another technique varies on considerations such as the availability of historical information and the complexity of creating an appropriate system model.

In addition, signal processing methods are necessary to examine and observe changes in infrastructure and asset performance and trigger an alarm if necessary. According to Liu et al., 2015 and Yang et al., 2016a, b conventionally, time and frequency domain methods are used in industrial condition maintenance practices. However, suppose environmental changes and other considerations, such as non-technical impacts, are considered. In that case, the condition of the infrastructure and asset could be assessed by comparing the infrastructure and asset performance with that of adjoining components or activities. However, (Liu et al., 2015) noted that condition-based maintenance practices efficiently identify irregularities in the infrastructure and asset performance; not every breakdown mode could be correctly described, and some could go unnoticed.

2.8 Infrastructure and Asset Maintenance Procedures

To design consistent infrastructure and asset maintenance procedures for effective operations and maintenance in all activities responsible for network operations. Consistent scope of the site specifics will lead the management to prompt outcomes on the action plan. Therefore, the required infrastructure and asset maintenance activities are planned preventive maintenance (PPM) – The strategy includes regular inspection, servicing and maintaining the infrastructure and assets in good condition (Kehinde et al., 2017; Thai et al., 2021). The maintenance performed at prearranged periods or cycles based on the specified benchmarks aims to lessen the possibility of failure or breakdown on the degradation of the working of the infrastructure and asset. The planned preventive maintenance should be applied to active and passive infrastructure and asset maintenance. The PPM is designed to develop a system that will uncover possible failures and make modifications or repairs to avoid failure.

This procedure also requires creating a technological and human procedure that will make the procedure operate within the acceptance level (Chen et al., 2021) and be used selectively in infrastructure assets and other equipment. Corrective maintenance (CM) – This maintenance procedure is constantly needed irrespective of the level of planned preventive maintenance used, but to a reduced degree (Kehinde et al., 2017; Duarte & Santiago, 2023). This procedure lets the infrastructure and asset operate until it gets a defect or failure that will not permit it to be performed again (Opara et al., 2021). The maintenance performed after fault detection aims to repair the fault to a stable condition immediately. These procedures are also os performed to rectify the failure or faults.

However, the most significant attention of maintenance practice should be a well-defined process that is precise and measurable. Failure to focus on the activities that constitute the maintenance activities will impact the infrastructure and asset performance (Kehinde et al., 2017; Opara et al., 2021). This proposition explains the significant impact of infrastructure asset performance and asset management activities. Thus, the below discussions on the maintenance context and operational models for telecom infrastructure and asset operations in emerging telecom operations and environment.

2.9 Integration of Human-Machine Collaborative Maintenance

During the maintenance process design phase, the knowledge of skilled individuals is recorded, captured and codified in the knowledge artificial intelligence management platform. Integrating this knowledge with complete data, machine learning training and process design concepts are applied to the management platform of the iterative and deterministic jobs formerly conducted by maintenance team members and continuously improve (Opara et al., 2021). Additional tasks are slowly passed to the machine through repetition, enhancing the maintenance implementation and practice correctness. In the telecoms infrastructure and asset domain, the maintenance process will be automated, in addition to planned preventive maintenance activities, by implementing predictive and self-healing abilities into the artificial intelligence management platform using collected data.

The human-machine collaborative maintenance process will leverage and value traditional maintenance approaches by deploying the artificial intelligence knowledge management platform to perform maintenance implementation and integrate existing processes, interfaces and needs. However, this approach is not a substitute for human operations but rather an enabler for an individual to create more value with the help of the machine through artificial intelligence. For example, the existing field operations technicians' roles will be upgraded to network strategy technicians, while new positions will be designed to manage the decision-making and special activities.

3 Research Method

The research method for this paper is a case study with a survey strategy that includes a quantitative technique structured questionnaire, a qualitative approach and a systematic review of related literature and documents data collection. The justification for using a case study survey approach is to capture a variety of viewpoints quickly, economically, focused, scientifically and reliably and the opportunity to use mixed techniques. Yin (2014) concluded that a mixed-method of quantitative and qualitative case study method offers the opportunity to gain a greater understanding of a contemporary phenomenon, generate hypotheses and reduce the possibility of any bias (Kezar, 2002), thereby reducing the researcher and participants' positionality. Indeed, variables in the questionnaire are likely scales from their descriptions and are employed in relevant asset management research. The five-point Likert scale uses intervals ranging from 1 strongly disagree to 5 = strongly agree based on Newell and Goldsmith's (2001) work on the question that evaluates perceptions of the variables.

Additionally, Creswell and Poth (2016) suggest that the survey strategy is one of the most common methods in management research that is generally used to answer the where, who, how and what research questions. This understanding is based on deductive reasoning, which begins with a theory and attempts to agree or disagree through quantitative techniques, and inductive reasoning, that the assumptions are derived from a particular phenomenon through a qualitative approach. The study participants are the employees of the telecom organisations responsible for the managed services, network operators and regulatory agencies. The judgment and area sampling techniques were used to administer the survey. This technique involves identifying the population within the studied operating environment that conforms to the criteria of high density with a focus on telecoms asset management. All participants had various background knowledge and experiences in telecoms and maintenance operations, regulatory and stakeholder management.

The online structured questionnaire was created and distributed to participants involved in asset maintenance design and implementation. These questionnaire questions reflect the research issue. For instance, one of the key questions was. How does asset performance improve the quality of service in your network? This question focuses on a specific issue with detailed and remarkable answers. At the same time, the semi-structured qualitative method was administrated to the managers in charge of the operations strategies, decision-making and stakeholder management. The qualitative questions focus on the participant's experience, knowledge and views concerning the research problems. For instance, How has the quality of services been affected by asset outages in your network? These questions were designed and structured to generate conclusive and quantifiable data.

The data were then analysed to understand the asset management process and how decisions are made in the physical context where the outages occur and affect QoS. The paper also reviewed related published journals on telecoms and asset management maintenance. This search includes keywords such as passive infrastructure and asset, network availability, QoS, asset management and preventive and corrective maintenance. This document search was conducted in vital academic databases such as Google Scholar, ProQuest, Scopus, Science Direct and JSTOR.

4 Results

As mentioned earlier in this paper, little attention has been given in the literature to explore the impact of different organisational scopes on asset management, performance, cost and risk in the research context. Therefore, this paper explores the positive effect of asset performance, cost and risk on QoS. Accordingly, various subjects were conceptualised to design the questionnaire, with each question examined on the five-point Likert scale of 1 – strongly disagree and 5 – strongly agree.

4.1 Quantitative Structured Questionnaire Results

The structured questionnaire was administered to participants at the researched organisations with managed services companies, with a returned response of 85% of the one hundred fifteen billed participants. This sample size is within the accepted requirements for the SPSS dataset (Kline, 2005). This population show adequate content validity based on Gable and Wolf's (1993) suggestion that the number of professionals needed for content validity is between two and twenty.

4.2 Results -Descriptive Analysis

Table 1 presents how the appropriate regression model fits the dataset. This is because .963 illustrates an appropriate linear correlation between the predictors and response variables. The coefficient of determination, which is R-squared 0.928, is a perfect proportion of the variance in the response (dependent) variable that is described by the predictor (independent) variables.

For this paper model, the R-square value signifies that the predictor (independent) variables would describe 93% of the variance in the response (dependent) variable. Field (2018) noted that the R-squared value is always lower than the R-squared value. The observed standard error of the estimated value of this model falls an average of 0.360945239 units from the regression line (Field, 2018); this is because the standard error of the estimate of regression is the average distance that the observed values fall from the regression line.

The F-test shows a p-value of <.001, which determines the significance of the R-squared change. This F-test is within the accepted value of 0.05 (Field, 2018), thus showing a significant relationship between the predictor (independent) variables and response (dependent) variable based on the p-value of <.001.

Table 1 Model summary

Model	R	R square	Adjusted R square	Std. Error of the estimate	Change statistics				
					R square change	F change	df1	df2	Sig. F change
1	.963[a]	0.928	0.921	0.360945239	0.928	129.208	9	90	<.001

[a]Predictors: (Constant), MTTR, Dieselmgt, Tower mce, PdM/CbM, Sparemgt, RCA, Janitorial, Asset Pef, PPM

Table 2 ANOVA[a]

Model		Sum of squares	df	Mean square	F	Sig.
1	Regression	151.501	9	16.833	129.208	<.001[b]
	Residual	11.725	90	0.13		
	Total	163.227	99			

[a]Dependent Variable: QoS
[b]Predictors: (Constant), MTTR, Dieselmgt, Tower mce, PdM/CbM, Sparemgt, RCA, Janitorial, Asset Pef, PPM

4.3 Results of the Overall Regression Model

Table 2 The sig value in the ANOVA tables provides the statistical significance of the regression model through the interpretation of each parameter value. For instance, the regression degree of freedom number is equal to the value of the regression coefficient minus one. This model intercept term is nine (9) predictors (independent) variables. With a total regression coefficient of ten (10), the regression of freedom is 10–1 = 9. In addition, the total degree of freedom which is the number of dataset (observations) participants minus 1. Therefore, 100–1 = 99. Also, the residual degree of freedom, which is the value equal to the total degree of freedom df minus regress df, 99–90 = 9.

On the other hand, the means square is analysed by the regression value of the square divided by the regression degrees of freedom df, which is 151.501/9 = 16.833344. Results from these values and parameters show a strong relationship between the predictors (independent) variables and response (dependent) variable. These indicators are in agreement with (Field, 2013; Blume et al., 2019) assertions on evidence to assume that the regression model fits the dataset appropriately.

4.4 Results from the Hypothesis Testing

Table 3 indicates a strong correlation between quality of service (QoS) with asset performance caused by maintenance activities of planned preventive maintenance with p-v of <.00, diesel mismanagement with p-v of 0.005, incorrect root cause analysis with p-v of 0.007 and means-time-to-repairs (MTTR) with p-v of 0.003. These predictors impact asset management and maintenance performance and, in turn, are factors impacting the quality of service.

Conversely, we observed a significant correlation between tower maintenance and janitorial with a p-v of 0.035. Although, these two predictors are non-traffic affecting and do not impact the quality of services. Another non-traffic affecting factor is the correlation between spare management and janitorial maintenance with a p-v .022. These non-affecting factors could only impact the quality of service at a secondary level of maintenance activities when a planned network comes with a work order and reference number to enable the implementation of the activity.

Table 3 Coefficient[a]

Model	Unstandardized coefficients			Standardized coefficients	t	Sig.	95.0% confidence interval for B	
		B	Std. Error	Beta			Lower bound	Upper bound
1	(Constant)	−7.918	1.064		−7.443	<.001	−10.031	−5.804
	Tower mce	0.012	0.078	0.005	0.153	0.879	−0.144	0.168
	PdM/CbM	−0.033	0.079	−0.013	−0.417	0.678	−0.19	0.124
	Janitorial	0.17	0.087	0.061	1.952	0.054	−0.003	0.343
	PPM	0.987	0.177	0.581	5.564	<.001	0.634	1.339
	Asset Pef	4.178	0.857	0.399	4.878	<.001	2.476	5.88
	Sparemgt	−0.061	0.083	−0.024	−0.735	0.464	−0.226	0.104
	RCA	−0.208	0.202	−0.077	−1.03	0.306	−0.61	0.193
	Dieselmgt	2.639	0.955	0.135	2.762	0.007	0.741	4.536
	MTTR	1.59	0.549	0.098	2.897	0.005	0.5	2.681

[a]Dependent Variable: QoS

In addition, the strong correlation from mean-time-to-repairs (MTTR) with a p-v of .003 as a factor affecting the quality of service shows that intermittent outages of infrastructure and assets impact network availability and QoS and increased OPEX caused by inappropriate planned preventive maintenance activities and diesel mismanagement, which have shown a significant correlation of beta 0.581, p-v of <.001 and beta 0.135, p-v of 0.007. In most cases, the generator elements and engine oil are not replaced at the appropriate cycle, and air-conditioning (cooling systems) units are not serviced according to the design specifications. The reason has been inadequate real-time solutions, such as predictive-based maintenance strategies to monitor maintenance activities and functionality. Another reason is based on the manual delivery of diesel supply to sites and inaccurate recording of quantity consumed and delivered based on the power systems counter issues.

We observed from the regression analysis that infrastructure and asset failures caused by diesel mismanagement, resulting from incorrect diesel allocation, quality and quantity consumption, poor monitoring of diesel supplied to sites, faulty generator counters and no intelligent system to confirm actual supply and consumption impact asset management and maintenance which in turn affect network quality of services. This concern about poor diesel management and the exact quantity consumed creates an increase in OPEX and outages that affect QoS and overall performance.

Root cause analysis with beta 0.135 and p-v of 0.007 shows a positive correlation to asset management and maintenance performance based on the invisibility of actual asset outages, which in turn increase frequent site visits by field technicians from the point of high mean-time-to-repairs (MTTR), penalties for not meeting the service level agreement (SLAs) and travel time, and off-course operational risks to personnel and the organisation. By risk, the paper focuses on the MTTR caused by the downtime from the infrastructure and asset failures at midnight or odd hours and not only the risk caused by the improper planned maintenance of the infrastructures and assets. Therefore, the result was based on addressing the impact of infrastructure and asset performance, OPEX and risk on QoS. The factors concerning performance were created based on the participants' response and their reviewed faults reports and documents, as additionally, the OPEX and risk were mainly from the available records and reports.

The predictive-based and condition-based maintenance approaches are another critical factor but with a negative correlation with beta −0.013, p-v 0.678 that affect the quality of service. This factor entails monitoring asset functionalities such as spare replacement, fault detections, real-time escalation on asset unforeseen breakdowns and outages and cooling systems performance. Inefficient cooling causes high temperatures, which in turn shut down the power systems and damage the active components because the active components operate within a specific temperature. Real-time escalation and detection of unforeseen asset outages reduce MTTR, improve performance and QoS, and reduce OPEX and security and environmental issues.

For instance, the results show intermittent asset breakdown, which impacts network availability and QoS and increases OPEX. The reason for the increase in OPEX has been frequent site visits by field technicians from the point of high mean-time-to-repairs (MTTR), penalties for not meeting the service level agreement (SLAs) and travel time, and off-course operational risks to personnel and the organisation. By risk, the paper focuses on the MTTR caused by the downtime from the asset failures and not only the risk caused by the improper planned maintenance of the infrastructures and assets. Maintenance activities of planned preventive maintenance represent the monthly asset maintenance cycle based on the manufacturer manual, which occurs every 250 h, 500 h and 1000 h based on the power configuration and hybrid solutions. Inappropriate and non-compliance to planned preventive maintenance activities have been observed from the data and is the key factor impacting asset performance. Reasons for non-compliance are basically human attitude and behaviour issues. Diesel consumption and management are other critical factors that impact asset performance. The data indicates inaccurate diesel consumption values, under allocation and quality issues. However, because of the human interface, there is concern about exposure to theft based on the demand on the local and street market, thereby creating concerns about diesel management.

4.5 Qualitative – Semi-Structured Interview Results

The paper also performs six (6) semi-structured interviews with participants responsible for policy making, supervision, asset management and maintenance decisions. This interview process was via Zoom and Google Meet and lasted around 30 min maximum duration. The results from the interview were a follow-up to the quantitative data collection that needed further explanation from management team members. The sampling strategy for this paper was purposeful sampling to assist with participants that are most effective and meaningful in answering the research questions (Lincoln & Guba, 1985). The participant was selected based on their length of work experience, understanding of the maintenance practice, readiness to share thc experience and how assets are managed in the organisation.

The collection of data through the interviews and survey questionnaires is simultaneous with the data interpretation and analysis to show a rich understanding of the data. Together with the textual data, this paper uses NVivo version 12 to analyse the data by organising, analysing and visualising the data through word search to code sources and capture concepts for thematic analysis. The outcome from the interview data shows that predictive-based maintenance strategies are significant in addressing issues of asset performance, which, in turn, impact the quality of service.

Figure 4 The word cloud evidently articulates each keyword or theme identified from the data and gradually organises comparable themes into various all-encompassing proportions or scopes that made up the foundation of a developing thematic structure that assisted in showing the evidence of the results. The thematic structure of the asset performance with the maintenance activities that are implemented through the predictive-based maintenance approach explains the stability of the network's quality of service. The asset outages depend on real-time escalation monitoring and functionality and how organisations in this context will use predictive-based maintenance strategies to improve asset performance and reduce operating expenditures. Keywords that relate to asset performance which impact the quality of service from the interviews are real-time, outages, functionality, monitoring and operating expenditure OPEX, etc.

Additionally, the results from the interview word cloud integrated and correlated with the results from the quantitative data. This insight provides directions on the impact of asset performance on QoS based on its maintenance activities. For instance, from the transcribed text from the interviews based on participants' responses, one participant's statement response to one of the questions on factors impacting asset performance made this proposition; "The issues with asset

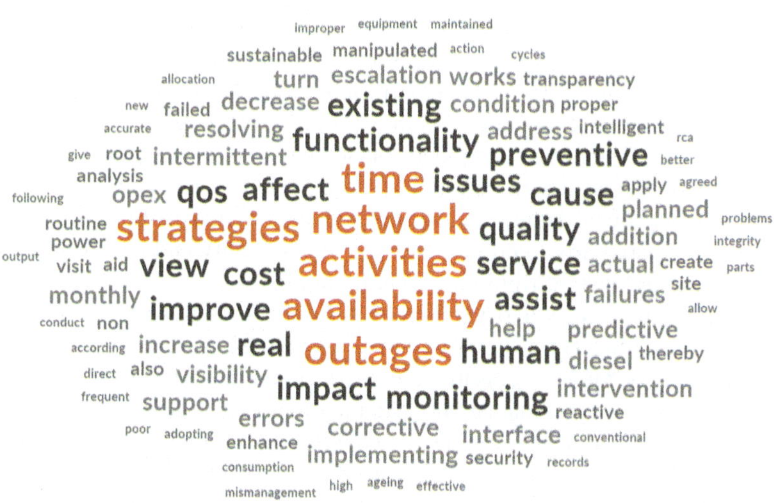

Fig. 4 Results from interview word cloud

performance is based on high mean-time-to-repair which is a result of the poor escalation of outages and sites location." This kind of issue could be resolved with the implementation of a predictive-based maintenance strategy whereby asset conditions and functionality are inspected before failures occur.

Other participants said that; "Increased visibility and transparency of all planned preventive maintenance activities to ensure that the servicing materials are actually replaced every cycle to avoid intermittent asset outages." These propositions from the participants are in alignment with various theories on maintenance strategies that address asset outages, such as using predictive-based systems to monitor maintenance activities and reporting field team activities.

4.6 Systematic Review Results

The results from the reviewed literature show the relationship between factors affecting the quality of the network from the asset management perspectives, such as the maintenance strategy, planning, optimisation approaches and asset maintenance activities and tasks (Dekker, 1996; De Jonge & Scarf, 2020). Therefore, a detailed assessment and categorisation of maintenance procedures are necessary for maintenance strategy (Wang, 2002). A critical categorisation of maintenance strategies considers three principal areas (Ayu & Yunusa-Kaltungo, 2020): planned preventive, corrective, condition-based and predictive-based maintenance strategy. Planned preventive maintenance prevents unplanned outages through programmed periodic inspections and spare replacement.

Critical activities in planned preventive maintenance involve engine oil or lubricant changes, filter or elements replacement and adjustment of belts and general inspections. These activities are repeated at intervals based on the manufacturer's references and analysis of quality factors such as mean-time-to-failure (MTTF). This preventive maintenance strategy guarantees good asset condition and functionality that reduces the risk of likely breakdowns or outages.

In contrast, Mobley (2003), Farinha (2020), and Duarte and Santiago (2023) noted that a preventive maintenance strategy does protect against unaccepted outages and defects of elements, as replacing spare parts too often is not always a better option based on the argument that new spare parts are more likely to be defective than the existing ones. However, the paper does agree with this proposition of replacing new spare parts to protect the asset functionality and performance based on experiences and insight from the extant literature.

Therefore, using the Nvivo version 12 software to analyse the imported data from the reviewed literature shows key findings that relate to the research, such as predictive data, systems, digitisation, twin, machining and systems, etc. These words are relevant to asset management and maintenance practices (Fig. 5).

Typically, if these strategies are integrated, planned preventive strategies for well-recognised and constant failure–time correlation, corrective strategies for assets with low criticality in terms of infrastructure and asset availability and cost, and condition-based strategies for the most critical assets (Molęda et al., 2023). In another category, predictive and proactive maintenance are recognised (Lawrence & O'Connor, 1995;

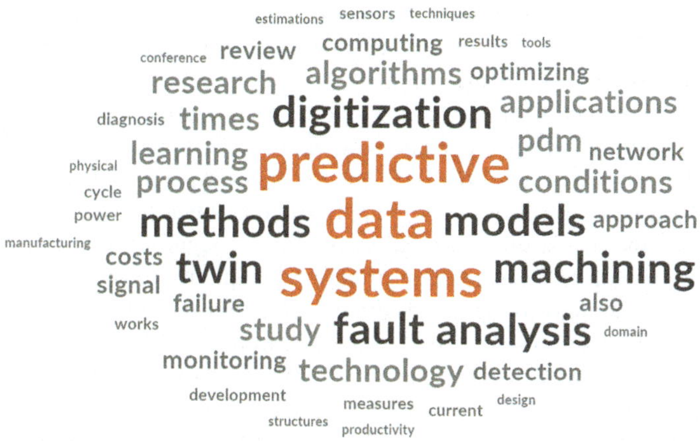

Fig. 5 Results from word cloud analysis

Van Horenbeek & Pintelon, 2013; Opara et al., 2021). These approaches are considered model-driven or condition-based maintenance, respectively.

Though both intend to improve the asset lifecycle and infrastructure and asset availability, the first uses historical data, and the second depends on continuous monitoring to identify early indications or symptoms of infrastructure and asset failure. Sikorska et al. (2011) analyse the methods of maintaining asset useful life by categorising the features of the methods in the context of resources and customer needs. Diagnosis that includes a description of methods and application in areas such as fault detection and identification, pattern recognition and root cause analysis. Gao et al. (2015) focus on showing fault detection and identification and labelling them in detail, grouping them as model-based and knowledge-based with data-driven (Moleda et al., 2020).

Additionally, Solé et al. (2017) argued on the root cause analysis issues concerning maintenance requirement, performance and scalability aspects. A prediction that covers methods and applications such as predictive management and useful lifecycle mainly in predictive maintenance application in broad industrial domains, such as Carvalho et al. (2019), focus on approaches, devices and data sources (Diez-Olivan et al., 2019) on the categorisation into descriptive, predictive and prescriptive analysis, (Zonta et al., 2020) on the limitation of predictive-based maintenance. The prescription describes advanced analysis applications in the prescriptive area, including techniques such as digital twin, simulation, process optimisation and area that is telecoms domain and other industrial areas covered by the review.

4.7 Summary of the Main Results

The summary of the key findings was outlined accordingly. First, to understand how network availability and QoS are affected by asset outages and performance, OPEX and risk. Secondly, the data shows that intermittent outages of the assets affect the network quality of services despite the planned maintenance cycles and diesel management of the assets. Other factors affecting network service quality are ageing infrastructure, assets, and security threats.

Table 4 presents the summary of the leading factors affecting the quality of services from asset performance perspectives. Asset performance with a strong correlation of .916 is significant at the 0.01 level, 2-tailed with p-v of <.001 as a factor affecting the quality of service. The inappropriate planned preventive maintenance activity with another strong correlation of .901 is significant at the 0.01 level, 2-tailed with p-v of <.001 as a factor affecting the quality of service. False escalation of root cause analysis with a positive correlation of .799 is significant at the 0.01 level, 2-tailed with p-v of <.001 as a factor impacting the quality of service.

Diesel mismanagement with a positive correlation of .519 is significant at the 0.01 level, 2-tailed with p-v < .001, affects the quality of service and means-time-to-repair (MTTR) with a correlation of .293 is significant at the 0.01 level, 2-tailed with p-v .003, is a critical factor affecting the quality of service. These factors are inconsistent with (Opata, 2013a, b; Opara et al., 2021; Chen et al., 2021; Thai et al., 2021; Danbatta & Zangina, 2022; Duarte & Santiago, 2023) studies on asset management and quality of services.

Additionally, the planned preventive maintenance activity to asset management and These identified factors impact the quality of the network services based on these reasons: asset failures caused by diesel mismanagement, resulting from incorrect diesel allocation, quality and quantity consumption, poor monitoring of diesel supplied to sites, faulty generator counters and no intelligent system to confirm actual supply and consumption.

These concerns about poor diesel management and the exact quantity consumed create an increase in OPEX and outages that affect QoS and overall performance.

The paper also found factors such as a lack of real-time visibility of the technical and environmental activities that affect QoS, from the point of not predicting the outages before they occur and resolving issues remotely before escalating to the field technicians to visit the site physically. Furthermore, not monitoring the maintenance activities remotely to ensure materials were appropriately replaced, and actual diesel quantity delivered to the site were also factors that affected QoS through port network availability.

5 Discussion

This paper supports the assumptions establishing intelligent and human collaboration by telecom operators in emerging operating environments and markets. This action will benefit them by executing appropriate asset management and maintenance practices through real-time applications that will address performance, cost and risk issues in their operations. Several approaches optimise the maintenance strategies for asset management support.

Table 4 Correlation

		QoS	Tower mce	PdM/CbM	Janitorial	PPM	Asset Pef	Sparemgt	RCA	Dieselmgt	MTTR
QoS	Pearson Correlation	1	−0.019	0.162	0.035	**.901****	**.916****	−0.102	**.799****	**.519****	**.293****
	Sig. (2-tailed)		0.852	0.108	0.733	**<.001**	**<.001**	0.311	**<.001**	**<.001**	**0.003**
	N	100	100	100	100	100	100	100	100	100	100
Tower mce	Pearson Correlation	−0.019	1	0.094	**−.211***	−0.036	−0.008	**−.244***	−0.002	0.06	0.006
	Sig. (2-tailed)	0.852		0.354	**0.035**	0.721	0.941	**0.014**	0.987	0.555	0.952
	N	100	100	100	100	100	100	100	100	100	100
PdM/CbM	Pearson Correlation	0.162	0.094	1	−0.181	0.167	0.182	−0.121	0.164	0.193	−0.01
	Sig. (2-tailed)	0.108	0.354		0.071	0.097	0.069	0.231	0.104	0.055	0.919
	N	100	100	100	100	100	100	100	100	100	100
Janitorial	Pearson Correlation	0.035	**−.211***	−0.181	1	−0.059	−0.017	−.228*	−0.051	0.066	−0.053
	Sig. (2-tailed)	0.733	**0.035**	0.071		0.557	0.865	0.022	0.616	0.511	0.601
	N	100	100	100	100	100	100	100	100	100	100
PPM	Pearson Correlation	**.901****	−0.036	0.167	−0.059	1	**.828****	−0.073	**.891****	**.266****	**.270****
	Sig. (2-tailed)	**<.001**	0.721	0.097	0.557		**<.001**	0.473	**<.001**	**0.007**	**0.007**
	N	100	100	100	100	100	100	100	100	100	100
Asset Pef	Pearson Correlation	**.916****	−0.008	0.182	−0.017	**.828****	1	−0.095	**.771****	**.632****	0.119
	Sig. (2-tailed)	**<.001**	0.941	0.069	0.865	**<.001**		0.345	**<.001**	**<.001**	0.239
	N	100	100	100	100	100	100	100	100	100	100
Sparemgt	Pearson Correlation	−0.102	**−.244***	−0.121	**−.228***	−0.073	−0.095	1	−0.189	−0.077	0.113
	Sig. (2-tailed)	0.311	**0.014**	0.231	**0.022**	0.473	0.345		0.06	0.446	0.262
	N	100	100	100	100	100	100	100	100	100	100
RCA	Pearson Correlation	.799**	−0.002	0.164	−0.051	**.891****	**.771****	−0.189	1	**.341****	0.055
	Sig. (2-tailed)	<.001	0.987	0.104	0.616	**<.001**	**<.001**	0.06		**<.001**	0.589
	N	100	100	100	100	100	100	100	100	100	100

(continued)

Table 4 (continued)

		QoS	Tower mce	PdM/CbM	Janitorial	PPM	Asset Pef	Sparemgt	RCA	Dieselmgt	MTTR
Dieselmgt	Pearson Correlation	.519**	0.06	0.193	0.066	**.266****	**.632****	−0.077	**.341****	1	0.005
	Sig. (2-tailed)	<.001	0.555	0.055	0.511	**0.007**	**<.001**	0.446	**<.001**		0.957
	N	100	100	100	100	100	100	100	100	100	100
MTTR	Pearson Correlation	**.293****	0.006	−0.01	−0.053	**.270****	0.119	0.113	0.055	0.005	1
	Sig. (2-tailed)	**0.003**	0.952	0.919	0.601	**0.007**	0.239	0.262	0.589	0.957	
	N	100	100	100	100	100	100	100	100	100	100

** Correlation is significant at the 0.01 level (2-tailed)
* Correlation is significant at the 0.05 level (2-tailed)

The existing asset maintenance practice is reactive, time-consuming, and not responsive or intelligent enough to address network operations and maintenance challenges. Current operational and asset management activities rely greatly on human interventions, affecting OPEX. These routine and non-routine operations and maintenance costs and activities represent high OPEX. The level of the existing intelligence or digital in the study context is mainly low. For instance, the integration between various power redundancy procedures (public grids/diesel generators 1 or 2) is not swift and intelligent.

Due to the more complex operating environments, such as across rivers, standalone or road coverage sites, that trigger network availability issues such that manual integration increases mean-time-to-repair (MTTR). This high MTTR explains the problem with network availability, which affects QoS, as monitoring whether the spares are replaced or reused with the correct quantity and quality. These problems could be resolved by predictive-based systems that would assist network operators in having visibility, enhancing performance, and reducing OPEX and risk. This is because the management and maintenance practice influences asset performance, improving the QoS that addresses customer satisfaction and expectations.

5.1 Contribution from the Reviewed Literature

The examined extant literature is distinguished between journal articles that focus on a wider explanation of approaches in the context of asset management and maintenance domain as a whole and asset engineering management presenting different solutions. However, issues relating to network quality of service, asset solutions and likely changes to existing practices, specifically maintenance strategies based on the context of this paper, were not addressed in detail in the reviewed literature. Thus, this paper extends the range of these reviews from the existing conventional methods and approaches used in asset maintenance to cover solutions that encompass systematic capabilities and intelligence-based procedures, such as predictive-based maintenance approaches. This understanding assisted with approaches covering the domains of diagnostics, prediction and remedy in the context of telecom operating environments. However, the reviewed literature suggests simplified strategies generic to fault detection and identification (diagnosis) and remaining lifecycle (prognosis) by classifying the approaches into dimensions of data-driven and model-based. Thus, from preventive and corrective maintenance strategies to predictive maintenance strategies.

5.2 Contribution to Knowledge

This study contributes to infrastructure and asset management and maintenance practice by developing a better understanding of the interactions and relationship between infrastructure and asset management and maintenance approach and asset performance and functionality. The paper builds on the existing literature from these domains to develop an empirical review and insight into how the organisation's resources could promote and improve network quality of service through asset performance. Thus, offering empirical findings would add to the scholarly study and offer a practical understanding of asset management employees in telco maintenance organisations. The results achieved during this study are presented concerning the research problem.

6 Conclusions

Using the predictive-based maintenance strategy and condition-based maintenance practices to address the factors affecting the quality of service through the prediction and real-time monitoring of asset functionality provides a new likely approach to addressing the asset performance issues affecting the network quality of services in emerging telecom environments and markets. Specific use of the predictive-based maintenance strategy, where traditional approaches requiring inactive elements and human analysis are substituted by industry 4.0R digital twin and artificial intelligence inference based on existing practice, extensive digitisation permitting real-time condition-monitoring of asset functionality and response from operator interface.

The proposed approach is related to real-time fault detection, identification and escalation based on the likely possessed understanding, data resources, assets and human interface, which is consistent with (Chen et al., 2021) studies on using intelligent and digitalised systems for asset management. However, improved QoS cannot be achieved without the stability of the asset performance. Therefore, network operators should focus more on asset management and maintenance activities towards addressing factors affecting QoS by integrating intelligence and a human-centric approach to optimise network values as a gateway for the future in sustaining better QoS. This proposition aligns with Duarte and Santiago (2023) studies on maintenance practice and asset functionality and effectiveness, where organisations focus on overall equipment effectiveness.

Additionally, implementing intelligent and human-centric strategies in asset management and maintenance practices in this context helped network operators focus on addressing and stabilising network quality of service in the future. This conclusion explains that adopting intelligence and human-centric operational concepts entailed practical intelligence of the technical and non-technical procedures that predict real-time and remotely resolve issues. Besides, a multi-disciplinary method between intelligence and human-machine interface through monitoring and control was necessary to address the network quality of service caused by poor network availability due to asset failures caused by the technical and non-technical factors associated with asset performance.

From the data analysis, the action needed to address the two key factors that impact asset performance and, in turn, network quality of services is to adopt a predictive-based approach that can assist in monitoring all asset management and maintenance activities concerning proper maintenance services on the assets, accurate delivering, actual root cause analysis on outages and supply diesel quantity to each location, real-time escalation of asset functionality and outages. In addition, assist field technicians to improve their routine activities by reducing frequent visit visits, false alarms or fault escalations. The action will result in improved asset performance, which in turn improves network availability and quality of service that will address customer complaints.

7 Recommendations

For network operators to address and achieve these QoS problems, a predictive-based asset management and maintenance practice that involves an intelligent and human-centric approach should be adopted and implemented to address asset performance issues affecting network availability and quality of services. This approach will assist in continuous real-time fault identification, escalation, prediction, amplification, and evaluation to improve maintenance decision-making based on asset functionality. This is because asset management, efficient maintenance approaches, and appropriate maintenance activities are crucial predictors of quality of service. This perspective supports the significance of integrating an improved intelligent and human-centric asset management and maintenance practice that will reduce too much reliance on human interventions and reporting.

This paper is to be further developed by the principal author of this paper, focusing on Off-Grid Passive Telecoms Infrastructure in Emerging Market: Efficient and Cost-Reduced in Asset Management Solutions for Sustainable Network Value.

References

Abd-Elrahman, A. (2019). A review of telecommunications service quality dimensions. *Proceedings on Engineering Sciences, 1*, 533–546. https://doi.org/10.24874/PES01.02.052

Adame, B. (2021). *The Ethiopian telecom industry: Gaps and recommendations towards meaningful connectivity and a thriving digital ecosystem.* [Online] Available at: https://doi.org/10.1016/j.heliyon.2021.e08146. Accessed 23 Mar 2022.

Almossawi, M. M. (2012). Customer satisfaction in the mobile telecom industry in Bahrain: Antecedents and consequences. *International Journal of Marketing Studies, 4*(6), 139–156.

Aremu, O. O., Palau, A. S., Parlikad, A. K., Hyland-Wood, D., & McAree, P. R. (2018). Structuring data for intelligent predictive maintenance in asset management. *IFAC-Papers Online*, 51(11), 514–519. [Online] Available at: https://doi.org/10.1016/j.ifacol.2018.08.370. Accessed 20 Dec 2022.

Ayu, K., & Yunusa-Kaltungo, A. (2020). A holistic framework for supporting maintenance and asset management life cycle decisions for power systems. *Energies, 13*(8), 1937. https://doi.org/10.3390/en13081937. Accessed 23 Mar 2022.
Barad, M. (1996). Total quality in management. In M. Warner (Ed.), *International Encyclopaedia of business and management*. Thomson Business Press.
Blume, J. D., Greevy, R. A., Welty, V. F., Smith, J. R., & Dupont, W. D. (2019). An introduction to second-generation p-values. *The American Statistician, 73*(sup1), 157–167.
Bradbury, S., Carpizo, B., Gentzel, M., Horah, D., & Thibert, J. (2018). Digitally enabled reliability: Beyond predictive maintenance. Available at:. https://www.mckinsey.com/business-functions/operations/our-insights/digitallyenabled-reliability-beyond-predictive-maintenance#. Accessed 15 Oct 2022.
Brady, M. K., & Cronin, J. J. (2001). Some new thoughts on conceptualising perceived service quality: A hierarchical approach. *Journal Marketing, 65*, 34–49.
Camci, F. (2009). System maintenance scheduling with prognostics information using genetic algorithm.
Carvalho, T. P., Soares, F.A., Vita, R., Francisco, G. D. P., Basto, J. P., & Alcalá, S. G. (2019). A systematic literature review of machine learning methods applied to predictive maintenance. *Computers & Industrial Engineering*137:106024. Available at: https://doi.org/10.1016/j.cie.2019.106024. Accessed 20 Nov 2022.
Chen, J., Lim, C. P., Tan, K. H., Govindan, K., & Kumar, A. (2021). Artificial intelligence-based human-centric decision support framework: An application to predictive maintenance in asset management under pandemic environments. *Annals of Operations Research*. [online] Available at: https://doi.org/10.1007/s10479-021-04373-w. Accessed 15 Oct 2022.
Creswell, J. W., & Poth, C. N. (2016). *Qualitative inquiry and research design: Choosing among five approaches*. Sage Publications.
Danbatta, K, & Zangina, T. (2022). Performance analysis of telecommunications regulations in Nigeria: A quality of service approach. *Dutse Journal of Pure and Applied Sciences*. 8 1a. [Online] Available at: https://doi.org/10.4314/dujopas.v8i1a.17. Accessed 29 Nov 2022.
De Azevedo, C. (2019). *Asset management insights. Phases, practices, and value*. Industrial Press, Inc.
De Jonge, B., & Scarf, P. A. (2020). A Review on Maintenance Optimisation. *European Journal of Operational Research, 285*(3). Available at: https://doi.org/10.1016/j.ejor.2019.09.047. Accessed 23 Mar 2023.
de Oliveira, A. K. V., Aghaei, M., & Rüther, R. (2020). Aerial infrared thermography for low-cost and fast fault detection in utility-scale PV power plants. *Solar Energy, 211*, 712–724.
Dellermann, D., Ebel, P., Sollner, M., & Leimeister, J. (2019) Hybrid intelligence. *Business and Information Systems Engineering*, 61. 1–7. Available at: https://doi.org/10.10007/s12599-019-00595-2. Accessed 20 Aug 2022.
Diez-Olivan, A., Del Ser, J., Galar, D., & Sierra, B. (2019). Data fusion and machine learning for industrial prognosis: Trends and perspectives towards industry 4.0. *Information Fusion. 50*:92–111. [online] Available at: https://doi.org/10.1016/j.inffus.2018.10.005. Accessed 15 Oct 2022.
Dekker, R. (1996). Applications of maintenance optimization models: A review and analysis. *Reliability Engineering and System Safety, 51*, 229–240.
Duarte, A., & Santiago, S. (2023). Maintenance practices and overall equipment effectiveness: testing the moderating effect of training. *Journal of Quality in Maintenance Engineering,* 29 2, pp. 442–459. [Online] Available at: https://doi.org/10.1108/JQME-04-2021-0033. Accessed 11 July 2023.
Duggal, E., & Verna, H. (2013). Service quality: Construct comprehension and evolution over time. *Journal of Services Research*, 135.
Farinha, J. (2020). From the maintenance management until the physical asset management – A disruptive paradigm. *Global Journal of Engineering Sciences*. [Online] Available at: https://doi.org/10.33552/GJES.2020.06.000635. Accessed 12 Aug 2022.
Field, A. (2013). *Discovering statistics using IBM SPSS statistics* (4th ed.). SAGE Publications.
Field, A. P. (2018). *Discovering statistics using IBM SPSS statistics* (5th ed.). SAGE Publications.

Gable, R. K., & Wolf, J. W. (1993). *Instrument development in the affective domain: Measuring attitudes and values in corporate and school settings*. Kluwer Academic.

Gao, Z., Cecati, C., & Ding, S. X. (2015). A survey of fault diagnosis and fault-tolerant techniques—Part I: Fault diagnosis with model-based and signal-based approaches. *IEEE Transactions on Industrial Electronics* 2015;62:3757–3767. https://doi.org/10.1109/TIE.2015.2417501. Accessed 15 Oct 2022.

Garvin, D. A. (1987). Competing on the eight dimensions of quality. *Harvard Business Review, 65*(6), 101–109.

Garvin, D. A. (1988). *Managing quality*. The Free Press.

Ghobadian, A., Speller, S., & Jones, M. (1994). Service quality concepts and models. *International Journal of Quality and Reliability Management, 11*, 43–66.

Grönroos, C. (1984). A service quality model and its marketing implications. *European Journal of Marketing, 18*, 36–44.

Grönroos, C. (1998). Marketing services: The case of a missing product. *Journal of business and Industrial Marketing, 13*(4/5), 322–338.

Halawa, F., Abdul, M., & Mohammed, R. (2022). Applying machine learning for duplicate detection, throttling and prioritization of equipment commissioning audits at fulfilment network. *IIE Annual Conference. Proceedings, 1*.

Hameed, Z., Ahn, S. H., & Cho, Y. M. (2010). Practical aspects of a condition monitoring system for a wind turbine with emphasis on its design, system architecture, testing and installation. *Renewable Energy, 35*, 879–894.

Hedvall, M., & Paltschik, M. (1991). Intrinsic service quality determinants for pharmacy customers. *International Journal of Service Industry Management, 2*(2), 38–48. cited in collier, *the service/quality solution,* 171.

ITU. (2019). *Economic impact of broadband in LDCs, LLDCs and SIDS*. An Empirical Study.

Kamar, E. (2016). Directions in hybrid intelligence: Complementing AI systems with human intelligence. *Proceedings of the international joint conference on artificial intelligence.*

Kehinde, A., Lawan, S., Adunola, F., & Isaac, A. (2017). GSM quality of service performance in Abuja, *Nigeria. International Journal of Computer Science, Engineering and Applications (IJCSEA), 7*(3/4), 40.

Kezar, A. (2002). Reconstructing static images of leadership: An application of positionality theory. *Journal of Leadership Studies, 8*(3), 94–109.

Kim, B., Jang, S., Jung, S., Lee, B., Puig, A., & Lee, S. (2014). A moderated mediation model of planned happenstance skills, career engagement, career decision self-efficacy, and career decision certainty. *The Career Development Quarterly, 62*(1), 56–69.

Kim, H., Choi, J., & Nam, Y. (2022). Characteristic analysis of the built environment of ferry terminals: A case study of Mokpo, South Korea. *Sustainability, 14*(4), 2115.

King, N. (2004). Using interviews in qualitative research. In C. Cassell & G. Symon (Eds.), *Essential guide to qualitative methods in organizational research* (pp. 11–22). Sage.

Kang, J. C., Sun, L. P., Lu, Y., & Sobral, J. (2019). An opportunistic condition-based maintenance policy for offshore wind farm. In C. Guedes Soares (Ed.), *Advances in renewable energies offshore* (pp. 753–760). Taylor & Francis Group.

Kline, R. B. (2005). *Principles and practice of structural equation modelling* (2nd ed.). Guilford Press.

Kotler, P., Kartajaya, H., & Setiawan, I. (2019). Marketing 3.0: From products to customers to the human Spirit. https://doi.org/10.1007/978-981-10-7724-1_10

Levenburg, N. M., & Magal, S. R. (2005). Applying importance-performance analysis to evaluate e-business strategies among small firms. *e-Service Journal,* 3(3), 29–48. [online] Available at: https://doi.org/10.2979/esj.2004.3.3.29. Accessed 20 Nov 2022.

Lewis, R., & Booms, H. (1983). The marketing aspects of service quality. In L. Berry, G. Shostack, & G. Upah (Eds.), *Emerging perspectives on services marketing* (pp. 99–107). American Marketing.

Lawrence, M., & O'Connor, M. (1995). The anchor and adjustment heuristic in time-series forecasting. *Journal of Forecasting, 14*(5), 443–451. https://doi.org/10.1002/for.3980140504. Accessed 15 Oct 2022.

Lin, Y. C., Su, Y. C., Lo, N. H., Cheung, W. F., & Chen, Y. P. (2013). *Radio frequency identification from system to applications*. Intech Open. Application of Mobile RFID-Based Safety Inspection Management at Construction Jobsite.

Lincoln, Y. S., & Guba, E. G. (1985). *Naturalistic inquiry*. Beverly Hills.

Liu, Q., Dong, M., & Peng, Y. (2013). A dynamic predictive maintenance model considering spare parts inventory based on hidden semi-Markov model. *Proceedings of the Institution of Mechanical Engineers, Part C: Journal of Mechanical Engineering Science, 227*(9), 2090–2103.

Liu, H., Loh, P. C., & Blaabjerg, F. (2015). Sub-module short circuit fault diagnosis in modular multilevel converter based on wavelet transform and adaptive neuro fuzzy inference system. *Electric Power Components and Systems, 43*(8–10), 1080–1088.

Liu, H., Zhou, J., Zheng, Y., Jiang, W., & Zhang, Y. (2018). Fault diagnosis of rolling bearings with recurrent neural network-based autoencoders. *ISA Transactions, 77*, 167–178.

Maletiˇc, D., Maletiˇc, M., Al-Najjar, B., & Gomišˇcek, B. (2018). Development of a model linking physical asset management to sustainability performance: An empirical research. *Sustainability, 10*, 4759.

Marquez, A., Fernandez, J., Fernandez, P., & Lopez, A. (2020). Maintenance management through intelligent asset management platform (IAMP). *Emerging Factors, Key Impact Areas and Data Models. 13*, 3762. [Online] Available at: https://doi.org/10.3390/en13153762. Accessed 29 Nov 2022.

Mathur, A., Cavanaugh, K., Pattipati, K., Willett, P., & Galie, T. (2001). Reasoning and Modeling Systems in Diagnosis and Prognosis. *Proceedings of SPIE - The International Society for Optical Engineering, 4389*. https://doi.org/10.1117/12.434239

Mobley, R. K. (2003). *An introduction to predictive maintenance*. Else.

Moleda, M., Momot, A., & Mrozek, D. (2020). *Predictive maintenance of boiler feed water pumps. Sensors, 20*(2), 571.

Molęda, M., Małysiak-Mrozek, B., Ding, W., Sunderam, V., Mrozek, D., & Mrozek, D. (2023). From corrective to predictive maintenance—*A review of maintenance approaches for the power industry. Sensors, 23*(13), 5970.

Nakajima, S. (1988). *Introduction to TPM*. Productivity Press.

Newell, S. J., & Goldsmith, R. E. (2001). The development of a scale to measure perceived corporate credibility. *Journal of Business Research, 52*(3), 235–247.

Nurysh, N., Naghavi, N., & Yin-Fah, B. (2019). Study on factors affecting customer satisfaction in mobile telecommunication industry in Malaysia. *International Journal of Recent Technology and Engineering., 7*, 299–306.

Olde Keizer, M. C. A., Teunter, R. H., Veldman, J., & Babai, M. Z. (2018). Condition-based maintenance for systems with economic dependence and load sharing. *International Journal of Production Economics*, 319–327. https://doi.org/10.1016/j.ijpe.2017.10.030. Accessed 15 Oct 2022.

Opara, G., Nwaukwa, J., Uloko, F., & Oborindo, C. (2021). Predicting asset maintenance failure using supervised machine learning techniques. World Journal of Intelligent Research (WJIR) 11, 3. 28–34. [Online] Available at: https://doi.org/10.31871/WJIR.11.3.6. Accessed 12 Dec 2022.

Opata, C. (2013a). *The curious case of consumer dissatisfaction in Nigerian telecommunications sector*. [Online] Available at: https://doi.org/10.2139/ssrn.2356804. Accessed 24 July 2022.

Opata, C. B. (2013b). Sustainable development and rural access to telecommunications in Nigeria. In C. Onwumechili & I. S. Ndolo (Eds.), *Re-imagining development communication in Africa* (pp. 225–244). Lexington Books.

Oviroh, P. O., & Jen, T. (2018). The energy cost analysis of hybrid and diesel generators in powering selected base transceiver station locations in Nigeria. Energies11:1e20. [Online] Available at: https://doi.org/10.3390/en11030687. Accessed 14 July 2022.

Pais de Almeida, J. E, Raposo, H. N, Farinha, J. T, Cardoso, A. M, & Marques, P. A. (2021). Optimising the life cycle of physical assets through an integrated life cycle assessment method. *Energies*. 14(19):6128. [Online] Available at: https://doi.org/10.3390/en14196128. Accessed 20 July 2022.

Parasuraman, A., Zeithaml, V. A., & Berry, L. L. (1988). SERVQUAL: A multiple-item scale for measuring consumer perceptions of service quality. *J Retailing, 64*, 12–40.

Parasuraman, A., Zeithaml, V. A., & Berry, L. L. (2015). SERVQUAL: A multiple-item scale for measuring consumer perceptions of service quality. *J Retailing, 64*, 12–40.

Peng, X., Scott, R., Prybutok, V., & Sidorova, A. (2014). Product quality vs. service quality in the mobile industry: Is there a dominant driver of customer intention to switch providers? *Operations Management Research, 7*(3/4), 63–76.

Raj, J. S., & Basar, D. A. (2019). QoS optimization of energy efficient routing in IoT wireless sensor networks. *Journal of IoT in Social, Mobile, Analytics, and Cloud, 1*(1), 12–23.

Rasmekomen, N., & Parlikad, A. (2016). Condition-based maintenance of multi-component systems with degradation state-rate interactions. *Reliability Engineering and System Safety, 148*(1–10). https://doi.org/10.1016/j.ress.2015.11.010. Accessed 23 Mar 2022.

Rinaldi, G., Thies, P. R., Johanning, L., & Walker, R. T. (2016). A computational tool for the proactive management of offshore farms. In *Proceedings of the 2nd international conference offshore renewable energy* (Vol. 19–20, pp. 111–115).

Rinaldi, G., Thies, P., & Johanning, L. (2021). Current status and future trends in the operation and maintenance of offshore wind turbines: A review. *Energies, 14*(9), 2484.

Santos, J. (2003). E-service quality: A model of virtual service quality dimensions. *Managing Service Quality, 13*, 233–246.

Satish, K., & Anil, A. (2017). Enhancing customer satisfaction through total quality management practices—An empirical examination. *Total Quality Management & Business Excellence, 30*, 1–21. https://doi.org/10.1080/14783363.2017.1378572. Accessed 23 Mar 2022.

Saunders, M., Lewis, P., & Thornhill, A. (2016). *Business research for business students*. Pearson Education.

Shafiee, M. (2015a). Maintenance logistics organisation for offshore wind energy: Current progress and future perspectives. *Renewable Energy, 77*, 182–193.

Shafiee, M. (2015b). Maintenance strategy selection problem: An MCDM overview. *Journal of Quality in Maintenance Engineering*. 21. 378–402. [Online] Available at: https://10.1108/JQME-09-2013-0063. Accessed 15 Nov 2022.

Sheu, S., Liu, T., Zhang, Z., & Ke, J. (2015). Extended preventive replacement policy for a two-unit system subject to damage shocks. *International Journal of Production Research, 53*(15), 4614–4628. https://doi.org/10.1080/00207543.2015.1005250. Accessed 23 Mar 2022.

Sikorska, J., Hodkiewicz, M., & Ma, L. (2011). Prognostic modeling options for remaining useful life estimation by industry. Mechanical Systems and Signal Processing 2011;25:1803–1836. https://doi.org/10.1016/j.ymssp.2010.11.018. Accessed 20 Oct 2022.

Solé, M., Muntés-Mulero, V., Rana, A. I., & Estrada, G., (2017). Survey on models and techniques for root-cause analysis. *arXiv preprint arXiv:1701.08546*.

Strauss, A., & Corbin, J. (1998). *Basics of qualitative research* (2nd ed.). Sage.

Sugeng, I. S. (2016). Understanding & indicators of customer satisfaction according to experts. Available at: https://Etalasepustaka.Blogspot.com.

Sugeng, W., Istiyanto, K. M., & Ashari, A. (2015). The impact of QoS changes towards network performance, international journal computer networks and communications. *Security, 3*(3), 48–53.

Thai, D. Duong, D. Falch, M. Xuan, C., & Thu, T. (2021). Factors affecting the sustainability of telecentres in developing countries. Telecommunications Policy, Elsevier, Vol, 46(3). [Online] Available at: https://doi.org/10.1016/j.telpol.2021.102265. Accessed 15 Nov 2022.

Van der Wal, R. W. E., Pampallis, A., & Bond, C. (2002). Service quality in a cellular telecommunications company: A south African experience. *Managing Service Quality: An International Journal, 12*(5), 323–335.

Van Horenbeek, A., & Pintelon, L. (2013). A dynamic predictive maintenance policy for complex multi-component systems. *Reliability Engineering & System Safety, 120*, 39–50.

Van Horenbeek, A., Pintelon, L., & Muchiri, P. (2010). Maintenance optimization models and criteria. *International Journal of System Assurance Engineering and Management, 1*, 189–200. https://doi.org/10.1007/s13198-011-0045-x. Accessed 23 Mar 2022.

Vu, K., Payam, H., & Bohlin, E. (2020). ICT as a driver of economic growth: A survey of the literature and directions for future research. *Telecommunications Policy, 44*(2), 101922.

Wang, H. (2002). A survey of maintenance policies of deteriorating systems. *The European Journal of Operational Research* 139: 469, 489. [Online] Available at: https://doi.org/10.1016/S0377-2217(01)00197-7. Accessed 20 Dec 2022.

Wolf, E. J., Harrington, K. M., Clark, S. L., & Miller, M. W. (2013). Sample size requirements for structural equation models: An evaluation of power, bias, and solution propriety. Educational and Psychological Measurement, 73(6), 913–934. https://doi.org/10.1177/0013164413495237. Accessed 20 Aug 2023.

Yang, K., Yang, M., Lang, X., Lang, Z., & Xu, D. (2016a). May An adaptive dead-time compensation method based on predictive current control. In *2016 IEEE 8th international power electronics and motion control conference (IPEMC-ECCE Asia)* (pp. 121–125). IEEE.

Yang, M., Yu, B., Honda, T., Zubair, S., & Sharqawy, M. (2016b). A maintenance-focused approach to complex system design. *Artificial Intelligence for Engineering Design, Analysis and Manufacturing, 30*, 263–276. https://doi.org/10.1017/S089006041600024X. Accessed 23 Mar 2022.

Yin, R. K. (2014). *Case study research design and methods* (5th ed., p. 282). Sage.

Zhao, H, Xu, F., Liang, B., Zhang, J, & Song, P. (2019). A condition-based opportunistic maintenance strategy for multi-component system. *Structural Health Monitoring.* 18(1):270–283. [Online] Available at: https://doi.org/10.1177/1475921717751871 Accessed 03 July 2022.

Zonta, T., da Costa, C. A., da Rosa Righi, R., de Lima, M. J., da Trindade, E. S., & Li, G. P. (2020), Predictive maintenance in the industry 4.0: A systematic literature review, Computers & Industrial Engineering 150, 106889. [online] Available at: https://doi.org/10.1016/j.cie.2020.106889. Accessed 15 Oct 2022.

Further Reading

International Telecommunication Union (ITU) (2017). Key ICT indicators for developed and developing countries and the world. [Online] Available at: www.itu.int/en/ITU-D/Statistics/Pages/facts/default.aspx. Accessed 10 Apr 2022.

Charles Okeyia is a doctorate research scholar at the University of Wales Trinity Saint David and a PhD scholar in Engineering Management at the University of Lisboa, Portugal. His doctoral work focused on telecom infrastructure and asset management and maintenance practices. Okeyia has a great pursuit in infrastructure asset management and sustainability in the emerging market. He has over 20 years of experience in telecoms infrastructure and asset management maintenance operations and services. Prior to the doctoral scholarship, Okeyia worked for a number of telecommunication organizations, including MTN Communication Limited, Africa's largest mobile network operator and LM Ericsson Telecommunication Company in Sub-Saharan African Operations.

Nuno Marques Almeida is an Assistant Professor in the Department of Civil Engineering, Architecture and Georesources of the University of Lisbon (IST) and a member of the research unit Civil Engineering Research and Innovation for Sustainability (CERIS). His entrepreneurial, professional, research and academic activities are focused on protecting and deriving value from engineering assets through the assimilation of innovative management and technological solutions for critical infrastructure, buildings and advanced industrial facilities of both the public and private sectors, towards more resilient cities and societies. Nuno has published more than 100 refereed journal and conference papers on these topics. He has advised transdisciplinary research and consultancy projects in the construction, real estate, water, energy, rail and road sectors. He is the head of the Portuguese delegation in the ISO/TC251 Asset Management standardization committee and is a member of the International Society of Engineering Asset Management.

Part II
Technology and Innovation in Asset Management

A Conceptual Implementation Process for Smart Maintenance Technologies

San Giliyana, Antti Salonen, and Marcus Bengtsson

Abstract Industry 4.0 is usually presented as usage of technologies. Some of these play an important role in the development of smart maintenance technologies. However, although the subject of smart maintenance has been discussed for more than 10 years, the manufacturing industry still finds it challenging to implement smart maintenance technologies to add benefits to maintenance organizations in line with company's goals. This study presents a conceptual process for implementing smart maintenance technologies, challenges and enablers to consider when implementing, and benefits. This article is based on an analysis of empirical findings from seven large manufacturing companies in Sweden, previous maintenance research, and authors' three previous smart maintenance research articles. In the first article, the authors explored perspectives on smart maintenance technologies from 11 large companies within the manufacturing industry, while in the second one, perspectives on smart maintenance technologies from 15 manufacturing Small and medium-sized enterprises (SMEs) were presented. In the third and final one, the authors developed and presented a testbed for smart maintenance technologies.

S. Giliyana (✉)
Mälardalen University, Eskilstuna, Sweden

Mälardalen Industrial Technology Center, Eskilstuna, Sweden
e-mail: san.giliyana@mitc.se

A. Salonen
Mälardalen University, Eskilstuna, Sweden
e-mail: antti.salonen@mdu.se

M. Bengtsson
Mälardalen University, Eskilstuna, Sweden

Volvo Construction Equipment Operations, Eskilstuna, Sweden
e-mail: marcus.bengtsson@mdu.se

A. Crespo Márquez et al. (eds.), *Advances in Asset Management: Strategies, Technologies, and Industry Applications*, Engineering Asset Management Review 3, https://doi.org/10.1007/978-3-031-52391-5_3

1 Introduction

In Industry 4.0, production, Information Technology (IT), and the Internet are combined (Matt et al., 2020). According to Bajic et al. (2021), the aim of Industry 4.0 is to integrate the physical world of manufacturing processes with the cyber world. Tao et al. (2019) state that the cyber world consists of data analysis, apps, services, and decision-making, while the physical world consists of machines, real factory environments, material, people, and execution. Information and Communication Technologies (ICT), smart factories, and the development of internet and embedded system technologies are what Industry 4.0 deals with (Liu & Xu, 2017).

The nine technologies of Industry 4.0 are: (1) Industrial Internet of Things (IIoT), (2) Big Data and Analytics, (3) Augmented Reality (AR), (4) Simulation, (5) Autonomous Robots, (6) Additive Manufacturing (AM), (7) Cyber Security, (8) Cloud Computing, and (9) Horizontal and Vertical System Integration (Alcácer & Cruz-Machado, 2019). These technologies provide innovations that affect production systems including strategies, processes, machinery types, and maintenance (Frost et al., 2019). Globalization, new devices, and connectivity drive industries to improve their production lines' performance and efficiency to stay relevant (Achouch et al., 2022).

To keep the production systems stable, maintenance as a function is essential and Abidi et al. (2022) state that maintenance activities are crucial to increase the lifetime of the equipment. In addition, Abidi et al. (2022) state that maintenance is one of the domains of manufacturing that introduces Industry 4.0 technologies, and maintenance is one function that will be affected by implementing Industry 4.0 technologies (Bokrantz, 2017). Four generations of maintenance are presented by Moubray (1997). In the first one, the machines were run to failure, which is related to Corrective Maintenance. Systems for planning and control were implemented in the second one, which is related to Predetermined Maintenance. Condition Based Maintenance (CBM) was presented in the third maintenance generation. The nine technologies of Industry 4.0, as well as Artificial Intelligence (AI) and Cyber-Physical Systems (CPS), play an important role in maintenance as a part of Industry 4.0 (Silvestri et al., 2020; Lee et al., 2019; Kanawaday & Sane, 2017; Al-Najjar et al., 2018) and may possibly be seen as the fourth generation of maintenance.

Previous research presents several approaches for smart maintenance technologies (Singh et al., 2013; Al-Najjar et al., 2018; Cachada et al., 2018). However, the manufacturing industry still finds it challenging to implement smart maintenance technologies to add benefit to maintenance organizations in line with company's goals. Giliyana et al. (2022, 2023a), Silvestri et al. (2020) and James et al. (2022), state that further research is needed to support the manufacturing industry in the implementation of smart maintenance technologies. Moreover, Flores et al. (2018) performed a study that included responses from 76 individuals located in 25 different countries. It showed that only 17% of the companies had a fully developed strategy for implementing Industry 4.0 technologies. Organizational challenges, such as integration of all stakeholders, Original Equipment Manufacturers (OEMs), end-users and support providers, are identified by Badri et al. (2018).

Furthermore, Lundgren et al. (2022) state that further research is needed to support the manufacturing industry in implementing smart maintenance. They have investigated several hindering factors in the implementation of smart maintenance, such as leadership clarity, culture, and time and resources. Lundgren et al. (2021), have studied many challenges when implementing smart maintenance in digitalized manufacturing industry, such as technological challenges in System Integration, which is also stated by Kans and Galar (2017). Additionally, many managerial and technical challenges are identified by Bajic et al. (2021) and Rikalovic et al. (2021), such as poor data quality, too large dataset to manage, and competence.

A study where 15 Small and medium-sized enterprises (SMEs) within the manufacturing industry were included, performed by Giliyana et al., (2023a), showed that no smart maintenance technologies had been implemented in those companies. According to Moeuf et al. (2020), SMEs offer 67.1% of the jobs in private sector in Europe, which means that the implementation of Industry 4.0 technologies is essential to compete nationally and internationally.

According to Bokrantz et al. (2020), smart maintenance has four dimensions. The first one is data-driven decision-making, making maintenance decision based on data. The second one is human capital resources, competence development of maintenance employees. Internal integration is the third, integrating the maintenance function with other internal functions, and the last one is external integration, integrating maintenance function with external functions, such as machine suppliers.

This paper addresses smart maintenance technologies and is related to the dimension of data-driven decision-making. The primary aim is to develop and present a conceptual implementation process for smart maintenance technologies. This process will include not only the challenges and enablers that the manufacturing industry needs to consider during implementation but also the benefits derived from the use of such technologies.

2 Methodology

This paper is based on a case study. The empirical data is collected through semi-structured interviews and analyzed through a process for qualitative data analysis.

2.1 *Data Collection*

The data collection is mainly based on interviews in 2023. Säfsten and Gustavsson (2020) present several weaknesses of interviews. One is that a wrong respondent may provide misleading results related to low validity (Säfsten & Gustavsson, 2020). To address this, the researcher required that the chosen respondents have experience in and work with maintenance development. When collecting qualitative data, purposeful sampling is often used (Patton, 1990), to select settings, persons, or events to provide important information (Maxwell, 1996). Säfsten and Gustavsson (2020) have written about convenience sampling, which means that the researcher uses the available elements. The corresponding author is an industrial doctoral student at Mälardalen Industrial Technology Center (MITC), which is a laboratory and collaboration platform in Sweden. The empirical data was collected through semi-structured interviews with respondents from seven large companies within the manufacturing industry, which were available within MITC's network for manufacturing companies and are at the forefront of technological development in Sweden. Two are classified as lighthouse factories, see Table 1.

The semi-structured interview questions were (shortened): (1) How do you see the term "Smart Maintenance"?, (2) How do you see the term "Smart Maintenance Technologies"?, (3) What types of technologies have been implemented or tested in your maintenance processes, and in what context?, (4) Was a process followed during the implementation of smart maintenance technologies?, (5) Would a specific process be needed?, (6) What are the challenges with the technologies in your maintenance processes? When implementing., (7) What are the challenges with the technologies in your maintenance processes? When using., and (8) What benefits do the technologies add to your maintenance processes?

2.2 Data Analysis

When the empirical data is collected through interviews, Säfsten and Gustavsson (2020) state that qualitative data analysis process should be used. Säfsten and Gustavsson (2020) present a process for qualitative data analysis. First, the data is

Table 1 Case companies

	Site empl.	Maint. depart. Size	Type	Position	Date	Dura. (min)
A	1700	230	Automotive industry	Maintenance manager	03–16	49
B	600	50	Automotive industry	Two maintenance engineers	03–16	24
C	10,000	1000	Automotive industry	Maintenance analyst	03–28	28
D	1000	70	Automotive industry	Maintenance engineer	03–30	35
E, lighthouse	2000	100	Manuf. of bearings, seals, and lubrication systems	Maintenance manager	04–27	32
F	5000	400	Automotive industry	Maintenance engineer	05–22	39
G, lighthouse	250	21	Manuf. of tools and machining solutions	Maintenance manager	05–23	43

reduced by transcribing and coding. Second, the data is visualized using tables, charts, etc. Third, the conclusions are made by looking for patterns, explanations, creating clusters, making comparisons, and analyzing changes over time (Säfsten & Gustavsson, 2020).

In this paper, the recordings from the semi-structured interviews were transcribed and coded in Nvivo, a software for qualitative data analysis. The codes were: implemented smart maintenance technologies, implementation process, smart maintenance definition, smart maintenance technologies definition, challenges when implementing, challenges when using and benefits of using. Then, the data was visualized through two tables, one for implemented smart maintenance technologies and implementation process and one for challenges when implementing, challenges when using, and the benefits of using smart maintenance technologies. In the last step, a deeper understanding was made by looking for explanations and making comparisons between different technologies and case companies.

3 Smart Maintenance Technologies and Their Benefits

In this paper, smart maintenance technologies are limited to the nine technologies of Industry 4.0, AI, and CPS. Using IIoT, the physical object can be connected to the Internet (Silvestri et al., 2020), through different types of communication protocols, such as Message Queuing Telemetry Transport (MQTT) and Open Platform Communications Unified Architecture (OPC UA) (Silva et al., 2021), and collect data. For instance, machine components can be connected to the Internet and thereby collect maintenance-related data, such as vibration, pressure, and temperature (Amruthnath & Gupta, 2018).

Witkowski (2017) and Yin and Kaynak (2015) have defined Big Data and Analytics through 5 V: (1) Volume, the amount of data, (2) Variety, the variety of data, (3) Velocity, the speed of new data generation, (4) Value, the value of data, and (5) Veracity, quality of the data. Two benefits of Big Data and Analytics are advanced data analysis and real-time decision-making (Witkowski, 2017; Subramaniyan et al., 2018).

Predetermined Maintenance is applied based on the failure time data. The failure time data is based on experiences or OEM, such as every 1000 h or ten days (Ahmad & Kamaruddin, 2012a). Labib (2004) presents three reasons why the last-mentioned strategy is unsuitable for minimizing operation costs and maximizing machine performance. The first one is that every machine works in a different factory environment. The second one is that the machine designers lack maintenance experience. Lastly, the OEMs may have a hidden strategy to maximize spare parts replacement. Implementing IIoT and Big Data and Analytics will overcome the drawbacks of Predetermined Maintenance. When the machines are connected to the Internet through IIoT, maintenance data can be generated, supporting the maintenance planning and decision-making, which is related to Big Data and Analytics and machine learning (Silvestri et al., 2020; Soori et al., 2023; Bona et al., 2021; Lee et al., 2019).

Another Industry 4.0 technology that may support maintenance planning is Simulation, which is based on mathematical modeling and algorithms used for process optimization (Erboz, 2017) and the design of a production system (Chong et al., 2018). According to Goodall et al. (2019), using Simulation, the behavior of a production system can be predicted, and thereby maintenance activities may be based on the prediction and the data from the simulation model.

Cloud Computing plays an important role in the development of smart maintenance technologies, which is about data and platform sharing at an entire company (Erboz, 2017). Using Cloud Computing, the maintenance data may be shared between, for example, the maintenance and production department, with the operators who work close to the machines and are responsible for Autonomous Maintenance. Furthermore, Cloud Computing can provide a cloud-based Computerized Maintenance Management System (CMMS) that may be accessed when the maintenance workers are close to the machines, using different types of client media, such as laptops, netbooks, and smartphones (Chang et al., 2016). In addition, by assigning different system roles, operators can view Autonomous Maintenance activities on their smartphones while near the machines. Maintenance engineers may find modules for maintenance planning, root cause analysis, and spare parts inventory useful. Maintenance managers may, for instance, benefit from a dashboard that includes various types of diagrams and graphs to follow-up maintenance progresses, based on (Chang et al., 2016; Carnero & Novés, 2006).

According to Roy et al. (2016) AR, which is based on a real-time combination of 3D-virtual objects with a real environment (Figueiredo et al., 2014), is playing a significant role in the development of smart maintenance technologies, offering step-by-step guidance for diagnostics, inspection, and training (Chong et al., 2018), which reduce time for performing maintenance tasks as well as errors (Masoni et al., 2017). Masoni et al. (2017) have shown the benefits of remote maintenance using AR, which is about remotely involving of the maintenance expert in problem-solving, by sending pictures of the real situation, and then, the maintenance expert replies through symbols, sketches, or text, in real-time. Some of the benefits of remote maintenance, mentioned by Masoni et al. (2017), are the reduction of travel costs and downtime. Another benefit of AR within maintenance is digitalized instructions for Corrective Maintenance, Preventive Maintenance, and Predictive Maintenance (Eswaran et al., 2023).

AM is another technology that may enable efficient maintenance (Chong et al., 2018). In AM, a digital design, i.e., 3D-CAD, is converted to a physical object, layer by layer. According to Chong et al. (2018), to create knowledge about equipment and maintenance, 3D-CAD can be used.

According to Kour and Gondhi (2020), the definition of AI is a technique that can perform activities that mimic human behavior. AI is becoming a major technology for developing smart maintenance technologies, such as the analysis of maintenance sensor data (Lee et al., 2019). Machine learning is a part of AI, and it is about making a system learn by itself based on data (Kour & Gondhi, 2020). Supervised machine learning forecasts events based on labeled data (Kour & Gondhi, 2020). Unsupervised machine learning is based on unlabeled data. Reinforcement machine learning is about a machine's ability to predict its own behavior to maximize performance (Kour & Gondhi, 2020). CBM consists of condition monitoring and decision-making (Ahmad & Kamaruddin, 2012b). In the first step, condition monitoring, the health of the equipment is monitored based on data collection, such as vibration data (Ahmad & Kamaruddin, 2012b). In the second step, decision-making, the decision is made. Diagnosis and prognosis are two types of CBM analysis processes (Lewis & Edwards, 1997). Diagnosis is about finding the source of a failure, while prognosis is about predicting the occurrence of a failure (Lewis & Edwards, 1997). The labeled data in supervised machine learning is based on condition monitoring data that can be used for failure prediction, which boosts the prognosis process (Yuan & Liu, 2013). Prajapati and Ganesan (2013) state that CBM aims to make decisions based on data. Furthermore, Prajapati and Ganesan (2013) state that machine learning algorithms aim to make decisions based on sensor data, which has a one-to-one relation with the aim of CBM. Vibration measuring is a CBM technique and the most spread technique for rotating machines (Azevedo et al., 2016). Al-Najjar et al. (2018) state that the process of vibration analysis is done manually, and due to the demand for labor, knowledge, and experience, the machines are not monitored continuously. Moreover, lubrication reduces friction and increases the machine's life. For instance, bearings are lubricated by central lubrication or manually greased, unrelated to the bearing condition and the real need (Al-Najjar et al., 2018). To overcome the mentioned drawbacks of CBM, Al-Najjar et al. (2018) have investigated how CBM can be boosted using CPS. The concept developed by Al-Najjar et al. (2018), consists of four steps, (1) Data collection, such as vibration data, (2) Maintenance actions recommendation, (3) Automatic maintenance of specific actions, and (4) Report to the maintenance department regarding what maintenance actions, when and where to have to be done manually.

In addition, Giliyana et al. (2022) have investigated what types of smart maintenance technologies have been implemented in the manufacturing industry and in what context. Their study shows that IIoT is implemented for machine connection, real-time sensors for maintenance data collection, Big Data and Analytics for maintenance data collection and maintenance planning, machine learning for predictive maintenance, AM for making spare parts for older manufacturing machines, and AR for remote maintenance.

Moreover, Giliyana et al. (2022) have mentioned several added values with smart maintenance technologies, such as the reduction of unplanned stops, automated condition monitoring, tracking of process data, failure prediction, spare parts availability, and deeper process knowledge. Furthermore, the respondents involved in Giliyana et al. (2022) mentioned that thanks to the smart maintenance technologies, they perform correct maintenance actions in time, they have become more cost efficient, they have fewer production disturbances, they work with predictive maintenance instead of reactive maintenance, they have increased knowledge on how their equipment works and can follow the degradation progress, and they plan maintenance activities and make decisions based on data (Giliyana et al., 2022).

4 Challenges and Enablers of Smart Maintenance Technologies

Masood and Sonntag (2020) have investigated several general challenges when implementing Industry 4.0 technologies. The most common one is the training of the workforce. Then, support from experts, time to work with the new technologies, awareness of a large amount of technologies, and a large investment. Bajic et al. (2021) have investigated several challenges, such as that the technology is not mature, awareness of what kinds of Industry 4.0 technologies exist, large investments and uncertain returns, and lack of strategy. Other challenges Bajic et al. (2021) mentioned are poor quality of the collected data, too large datasets to manage, competence about Industry 4.0 technologies, and Cyber Security and data protection. Badri et al. (2018) have investigated several implementation challenges, such as the involvement of end-users and support providers. In this case, end-users can be maintenance engineers, mechanics, electricians, operators, and so on. Giliyana et al. (2022) have investigated several challenges when implementing smart maintenance technologies in the manufacturing industry, such as Start-up cost, Cyber Security, change management, the technology may not be available for older machines, resources, competence, know what to monitor, and know what type of data to collect.

Moreover, Giliyana, Bengtsson, and Salonen (2023a) have investigated smart maintenance technologies implementation challenges related to Knowledge, Time and resources, Cost, and Age of the machines, from fifteen SMEs within the manufacturing industry in Sweden. Knowledge, they have summarized as, (1) To know what kind of data to collect and measure, (2) Technical knowledge, competence, and expertise in senior positions, (3) Make the technologies work and make everyone understand and work after them, etc. The Time and resources category presents, (1) Resources to work with these technologies, (2) The time between implementing these technologies and benefits, (3) The maturity of the technologies, and (4) The machine manufacturers do not have the opportunity to offer these types of technologies for maintenance. The Cost category, they have summarized as, 1) Start-up costs, 2) Financial resources, etc. Older machines are presented in the category Age of the machines, since some of the respondents in their study mentioned that their machines are older, and these types of smart maintenance technologies may not be available for older machines (Giliyana, Bengtsson, & Salonen, 2023a).

4.1 Implementation

Lundgren et al. (2021) have presented a smart maintenance implementation process. This process consists of (shortened): (1) Benchmarking using a smart maintenance measurement instrument, providing employees an understanding of the four smart maintenance dimensions presented by Bokrantz et al. (2020). (2) Setting clear goal based on the company's main goal. (3) Setting strategic priority to ensure that the activities are done in the right direction. (4) Planning key activities to reach the desired goals. (5) Creating an action plan, and (6) Follow-up to ensure the impact on the company's main goal.

A six-step process for digitalization of the industrial maintenance is presented by Campos et al. (2020). The steps are (shortened): (1) Business case, a need is recognized. (2a) Business understanding, a detailed understand of the situation is gained. (2b) Data understanding, data collection and storage, and ICT requirements. (3) Requirements specification, ICT requirements are specified and developed. (4) Subsystem/component design. (5) Prototype, an ICT prototype is designed, and (6) Final product/service, the ICT application is finalized.

CBM is expected to play a dominant role in smart maintenance technologies (Al-Najjar et al., 2018). Therefore, the four-step CBM process presented by Bengtsson (2008) is of interest for this research paper. In the first step, Feasibility test, the question regarding whether CBM technologies are applicable should be answered. In the second step, Analysis and technical development, responsibilities are assigned, components, sub-systems and/or systems to be monitored are selected, and what, how and when to measure, are answered. Bengtsson (2008) has divided the third step, Implementation, into (1) Management and (2) Introduction. Management entails managers facilitating and supporting the implementation process. In the introduction part, the technologies are introduced for other departments. When the implementation is done, in the fourth step, Assessment, the result needs to be compared to the period before CBM technologies, including employee perception, calculations like investment cost and cost for lost production, and so on. Another thing to consider is continuous improvement (Bengtsson, 2008). A similar CBM implementation process is also presented by (Rastegari, 2017). Furthermore, a CBM implementation framework is presented by Ahmer et al. (2022) and a general CBM process is presented by Ahmad and Kamaruddin (2012b).

Additionally, Giliyana, Karlsson, et al. (2023b) have developed a testbed for smart maintenance technologies and the development process is presented in their study. Two steps in their process that will play an important role in this study are Team building and Pre-study. Smart maintenance technologies are characterized as cross-border technologies, necessitating the involvement of multiple departments, such as maintenance, IT, and production departments. Hence, in Team building, it is imperative to construct a cross-functional team that includes maintenance engineers, mechanics, electricians, software developers, maintenance managers, production managers, and operators with direct proximity to the machines and with the deepest knowledge about machines and processes. According to Masood and Sonntag (2020), one of the challenges when implementing new technologies is awareness of many technologies. Therefore, in Pre-study, what Industry 4.0 technologies exist and what kinds of technologies can be used in the development of smart maintenance technologies, should be clarified (Giliyana, Karlsson, et al., 2023b).

5 Empirical Findings

The respondents see smart maintenance as a support system to solve problems smarter and faster based on data. The respondents exemplified smart maintenance technologies by stating AR, smart sensors, machine learning, data analysis, IoT platforms, 3D-printer, Cyber Security, OPC UA, and MQTT. Table 2 shows the implemented smart maintenance technologies, and the implementation process.

Table 3 shows that all case companies have implemented some kinds of smart maintenance technologies. Moreover, Table 3 shows the benefits of implemented smart maintenance technologies, as well as challenges.

6 Analysis

The empirical findings show that the respondents see the term of smart maintenance as a support system and enabler, rather than a revolution. For example, at case company G, an application development software and QR codes are used to improve Autonomous Maintenance, and at the case company F and G, 3D printers are being used to reduce waiting time for spare parts. Also, previous research presents that smart maintenance is a kind of support system to improve Preventive and Corrective Maintenance (Eswaran et al., 2023; Soori et al., 2023).

Regarding the term smart maintenance technologies, the respondents mentioned AR, machine earning, data analysis, IoT, cloud solutions, 3D-printer, etc. The respondent from case company C, mentioned that smart maintenance technologies are not about a robot coming and fixing a machine. Furthermore, this respondent mentioned that smart maintenance technologies are about making maintenance repairmen and technicians' job easier and faster. At the case company F, the respondent mentioned that technologies should make work easier for employees, otherwise, they will not use the technologies.

The empirical findings show that all case companies have implemented some kinds of smart maintenance technologies, such as mobile applications, sensors, automatic vibration sensors, dashboards, AI algorithms, IoT platforms, and visualization software, such as Grafana. Previous research present that AR is becoming very important for smart maintenance (Chong et al., 2018; Masoni et al., 2017; Eswaran et al., 2023), but one answer, by the respondent from case company F, is that they purchased AR equipment for remote maintenance, but they could not see the value within the frame of maintenance. AR is implemented but not used, as mentioned by the respondent.

This study also shows that the case companies face several challenges when implementing and using smart maintenance technologies, although the subject of Industry 4.0 has been discussed since 2011. Case companies G and E are lighthouse factories, but the empirical findings show that they still face many challenges when implementing smart maintenance technologies, such as support from IT

Table 2 The implementation of smart maintenance technologies

	Implemented smart maintenance technologies	Implementation process
A	Mobile applications and sensors. Ongoing pilot projects. "*...I feel that we are quite mature both in terms of installing the equipment, using the equipment, and evaluating the data...*".	"*...technology roadmap...*" to identify needs. A "*...guideline...*" could be needed for implementing new technologies.
B	Automatic vibration sensors to replace bearings before breakdowns.	"*...it's defined in the project..., we have a business plan, and the maintenance has its part in the business plan...*". A specific process could be needed for implementing smart maintenance technologies in a standardized way.
C	Proof of concept before deciding. Sensors are connected to collect data and monitor. 1.5 million Swedish crowns are saved on one compressor in one year. When the profit is visible, they decided to work with larger cases. "*We have a project for an IoT platform right now for a new battery factory.*". Furthermore, there are dashboards, machine learning, big data and analytics, and emails that are automatically sent to the technicians when anomalies are detected by machine learning. Regarding root cause analysis, there are algorithms that have not yet been implemented. Therefore, they do the root cause analysis manually.	(1) connectivity, which is about deciding what machine to connect and with what communication standard, (2) data acquisition, what data to collect and with what sensors, (3) data pre-processing, what data to store in the database, (4) data visualization, visualization of the data on dashboards, (5) data analytics, data analysis using algorithms, and (6) automatic maintenance, maintenance orders are created automatically and the degradation can be followed before breakdown.
D	IoT platform and measurements in welding robots. A few robots are connected, and they can see trends and tendencies. One reason for implementing smart maintenance technologies is to understand and learn what is available and what they can get out of the technologies.	An organization works with new technologies and identifies where they can benefit from smart technologies to reduce disturbances, improve quality, or reduce costs, in cross-functional areas with production, production technology, and maintenance. One improvement is to have a standardized process for implementing new technologies.

(continued)

Table 2 (continued)

	Implemented smart maintenance technologies	Implementation process
E	Grafana software for data visualization, such as moisture meters to identify corrosion problems. Thermometers, vibration measurement equipment, and other types of logs. Experiment with hardening processes by connecting machines to the Swedish Meteorological and Hydrological Institute (SMHI) and detect when there is a certain percentage of lightning strikes in the region, which then stops the processes for about 4–5 h, to ensure that it stabilizes.	The implementation is done in the form of a project. There is no structured way of working. They try to demonstrate the benefits of smart maintenance to get management approval for the projects. The plan is to start a project in each sub-factory to identify losses and then demonstrate the benefits with the help of smart maintenance. A specific process or Gant chart could be needed when implementing smart maintenance technologies. A process is needed to follow to show where to start when implementing technologies for smart maintenance, to get the most benefit. A process also makes it easy to involve the culture, as not all people are for new technology, and shows what is needed in the beginning, such as training.
F	3D printer is fully implemented to manufacture spare parts to reduce waiting time for spare parts. The 3D printer is connected to the CMMS. A library has been built with standard items. They are working on machine connectivity. They want to standardize how machines are connected and how data is visualized so that a maintenance engineer can determine the criticality of their equipment and connect their machines themselves. They purchased AR equipment for remote assistance, but they couldn't see their value within the frame of maintenance. They are implemented but not used.	They don't have a clearly defined process. When they implemented the 3D printer, the maintenance engineers evaluated it themselves, purchased a small 3D printer, set up various use cases, and then made calculations based on that. Afterward, they scaled it up by taking real use cases from the operations and ensuring local anchoring, ownership, and competence. The most important thing is to have local anchoring when implementing new technologies. A specific process for implementing smart maintenance technologies is not needed, but some types of manufacturing strategy where all functions are involved, so that each one does not do it separately, production, quality, maintenance etc.
G	They have conducted a pilot project for autonomous maintenance since they had issues with autonomous maintenance not being performed. They built an application using Microsoft power apps software and placed QR codes on the machine's checkpoint, such as checking oil. When scanning the QR code with a phone, a title appears indicating what needs to be done. Clicking on the title provides brief instructions on how to perform the task. It worked, but they have not been able to establish a stable solution to scale it up, therefore they only have it implemented on one machine. A 3D printer is used to print spare parts to reduce waiting time for spare parts.	They don't have an implementation process. For instance, 3D printing has been driven by ideas. One of the reasons why scaling up doesn't work at the case company could be the lack of an implementation process. "*Some things need to be driven by enthusiasm, I believe, and then you can't maybe frame it too much. But when it comes to implementation, you need to have some structure to follow.*" at a small company, perhaps, a process is unnecessary, but in a large company, there are so many people involved, such as the IT department, and if there's no process, there are many obstacles along the way. In the early stages, creativity is needed, but when things need to be streamlined, a process is necessary.

Table 3 The benefits and challenges of smart maintenance technologies

	Benefits	Challenges when implementing	Challenges when using
A	*"...with data and information, we gain an advantage and we get a competitive...", "...more fact-based decisions...", "we will optimize our way of running production, we will not wear out the equipment as hard, we will have a better economic use of our equipment...", "...we will work more safely..."*	*"Competence...", "...resists new things...", "...individuals and groups need to go through a journey of change. But at the same time, we live with old hardware."*	*"Certain technologies are quickly accepted while other are slower."*
B	*"... to become more efficient in maintenance, so that fewer manual resources are needed, and so that it becomes more accurate when doing the right maintenance at the right time."*	Competence and cost. *"... convincing that this is worth the investment, because a lot is linked to the fact that we want a quick return on invested capital, but with these, it's like 3, 4, 5 years maybe at best before you see the real profit..."*. Defining the responsibility is a challenge, since smart maintenance technologies are cross-border technology and both maintenance, IT, and production need to be involved.	To have a follow-up to ensure that the technologies are used in the right way. One mistake is to set a level on a vibration sensor to replace a bearing when it alarms without doing a root cause analysis.

(continued)

Table 3 (continued)

	Benefits	Challenges when implementing	Challenges when using
C	They mainly work on urgent jobs and "*...progressive maintenance...*", they repair machines that break down, with not much focus on improvements. This leads to long production stoppages and stress among the staff. With smart maintenance, if it is possible to see when a machine is going to break down, they can plan and ensure that the correct spare parts are in stock, and it may take a week to get a motor to the comp any. With smart maintenance, the production department, maintenance and logistic can be connected to each other, to coordinate better. This coordination does not exist today. "*This will be a significant advantage that we can get from smart maintenance.*". "*It took a very long time for the company to accept smart maintenance, I think it took 3 years. Because the company is a large company, and they work in a very traditional way. It was very difficult to come and say that data can do the job better for those who have been working here for a very long time.*" thanks to research projects and proof of concepts, the benefits of smart maintenance, such as the savings on the compressor, became apparent, and the decision was made to start with smart maintenance.	Competence. It is difficult to use the technology, difficult to choose signals, and difficult to use data and analyze it. Therefore, the progress is slow. A large SCADA system exists and collects a lot of data. But there are challenges in the maintenance department, such as training and acceptance.	When using the smart maintenance technologies, they have a standard package that they deliver and show to employees. It's not certain that those who receive the standard package feel that it provides value. What they do at the case company is to update the standard package based on inputs from employees. This is a way to increase interest and motivate them to use the technology. There are different requirements from departments, such as maintenance and production. The challenge is to implement all the requirements that come in, but all requirements are taken in and prioritized.

(continued)

Table 3 (continued)

	Benefits	Challenges when implementing	Challenges when using
D	*"In some places, not everywhere, but in some places, we can save quite a bit of money by using smart technology and measuring, being able to predict disruptions and breakdowns so that we can plan a component exchange at a good time instead of experiencing a long unplanned shutdown."*	To find a suitable prioritization, how do we do it, and who does it? It is a problematic area to find time to work with these things. *"It's not just a matter of connecting a sensor and then the problem is solved."* One big problem is data analysis, finding any correlations, setting limits, and seeing what's right and what's outside the tolerance. Not dedicating people with the right skills to do the data analysis and decide where the data should be stored, and how should it be stored and visualized in a simple and good way. There are many steps to consider, from the sensor on an equipment, the communication to the IoT platform and converting data to the information that can be used to make decision on.	It's becoming more and more automated. When a sensor is connected and collecting data via OPC tags and has done a basic analysis, then it rolls much by itself. However, maintaining competence is always a challenge, even in the user stage.
E	A decision support system to put the right maintenance action. They are repairing a lot of things expensively, and they are replacing too many components, and sometimes when a machine is disassembled, they take the opportunity to change other components too, and that is "*… money out…*".	Lack of time, resources, and competence. Acceptance from the machine owner. The production manager does not want the machines to be shut down. Cost is not a problem, but the challenge is to shut down production to install sensors.	The challenges when using smart maintenance technologies is to industrialize the solutions. Other challenges are setting alarm levels to identify vibrations that require replacement and how many months until that point, and to trust the technologies.

(continued)

Table 3 (continued)

	Benefits	Challenges when implementing	Challenges when using
F	3D-printer that has provided several benefits, such as shorter waiting time to spare parts. Data-driven decisions and auto-generated work orders will result in significant benefits in the terms of availability and working time. *"If you don't implement new technologies, you won't be in the game", "if you are going to engage in real large-scale production, you have to do this."*	*"The largest challenge is the alignment between organizations"*. There are many standards and technical suppliers, and the challenge is to adhere to a standard. Proper preparation is important during implementation. *"The issue that can arise if you don't do the preparation properly is that you may have difficulty justifying the business."* Another mentioned challenge is too large data to manage.	There must be a genuine need from the business. One example is AR, where there wasn't a genuine need, and therefore, the technology is not being used today, even though this is available at the case company. Technologies should make work easier for employees, otherwise, they won't adopt new technologies.
G	They have issues with waiting time for spare parts since they have many machines, and it is impossible to have all the spare parts in stock. *"If we can 3D print things, for example, and have the things in 2 days instead of 4 weeks, it's a great opportunity."* when it comes to autonomous maintenance instructions, for instance, it is an opportunity to make it less dependent on individual personnel, both in Terms of how it's performed and if it's performed.	The challenges when implementing new technologies are to get support from the IT department, competence, and resources.	*"…a form of competence or an understanding that this exists and can be used for these things."*

department, competence, resources, and too many communications standard protocols to manage. Additional implementation-related challenges mentioned by the respondents at the other case companies were competence, cost, and lack of time, all in alignment with Giliyana et al. (2022) and (2023a).

Regarding the challenges when using smart maintenance technologies, the respondents mentioned, ensuring that the technologies are used correctly, acceptance, competence in a user stage, and the risk of replacing broken components based on the output of a technology without doing root cause analysis. The respondent at six of seven case companies mentioned that they do not follow any specific implementation process for implementing smart maintenance technologies. At the seventh one, case company C, a six-step process is followed, but the focus is only on the technical parts. For instance, the respondent at case company A mentioned that one improvement could be to have a guideline for implementing new technologies. The respondents at case company B mentioned that a specific process for implementing smart maintenance technologies is needed to work in a standardized way, which is also mentioned by the respondent from case company D. Furthermore, the respondent at case company E mentioned that a specific process or Gant chart could be needed when implementing smart maintenance technologies, to show where to start the implementation to get the most benefit and that a process also makes it easy to involve culture, since not all people are willing to work with new technologies. One more benefit of having a specific process for the implementation of smart maintenance technologies, mentioned by the respondent at case company E, is that a process makes it easy to see what is needed in the beginning, such as training. Moreover, the respondent at case company G mentioned that one reason why scaling up at their company does not work could be the lack of an implementation process, and the respondent at case company F mentioned that a strategic way of working is needed where all functions are involved, so that each one does not do it separately, such as production, quality, and maintenance department. Finally, the previous research shows that smart maintenance technologies are cross-border technologies, where several departments need to be involved in the implementation process, which is mentioned by several respondents, such as from case companies B and D.

7 A Conceptual Implementation Process

The empirical findings and previous research show that manufacturing companies still find it challenging to implement smart maintenance technologies to add benefits to the maintenance organizations in line with company's goals, although many smart maintenance implementation processes are presented in previous research, such as Lundgren et al. (2021) and Campos et al. (2020). Furthermore, Silvestri et al. (2020) state that from industry's point of view, there is no clear view of what steps an organization should take to implement smart maintenance technologies. Related to this, a study performed by Fraser et al. (2015), shows that out of 82 empirical papers, only three have direct practical links to the manufacturing industry. In addition, previous research presents many challenges when implementing smart maintenance technologies. However, the challenges are not organized into a structured implementation process. One example is the challenges identified by James et al. (2022), Giliyana et al. (2022) and (2023a), related to the implementation of smart maintenance technologies. Also, the challenges that are identified by Bajic et al. (2021) and Rikalovic et al. (2021). Furthermore, as mentioned in the Analysis section, several respondents mentioned that a specific process or a standardized way, is needed to implement smart maintenance technologies.

This research paper organizes the challenges and enablers into a conceptual implementation process for smart maintenance technologies, see Fig. 1. The challenges and enablers are based on previous research and empirical findings from seven large manufacturing companies involved in this study. Furthermore, the conceptual implementation process could possibly support the manufacturing industry in what steps to take to implement smart maintenance technologies.

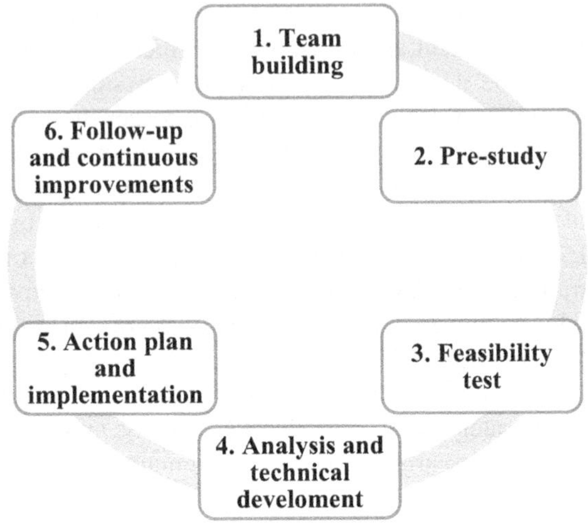

Fig. 1 The conceptual implementation processes

7.1 Team Building

Smart maintenance technologies are characterized as cross-border technologies. Therefore, in this step, a cross-functional team is built (Giliyana et al., 2023b). One enabler is good communication between functions (Bengtsson, 2008). As presented in the empirical findings, other enablers are using pilot projects and testbeds to demonstrate the benefits of smart maintenance technologies and get management approval. According to the empirical findings, some challenges to consider in this step are cost, lack of time and resources, resisting new technologies, and support from IT department.

7.2 Pre-study

In this step, smart maintenance technologies should be clarified. Therefore, one enabler for this step is education and training, since Masood and Sonntag (2020) have shown that knowledge is one major challenge when implementing new technologies. Additionally, Giliyana et al. (2022) and (2023a), have identified many challenges when implementing smart maintenance technologies. One major challenge is knowledge, about IIoT, Big Data and Analytics, AI, etc.

7.3 Feasibility Test

Based on the potential gains and drawbacks of the technologies the question of whether smart maintenance technologies are applicable should be answered in this step. Moreover, whether smart maintenance technologies are accepted in the company should be investigated in this step. One enabler is to use audit and benchmarking to assess the maturity of the maintenance organization (Bengtsson, 2008). According to the respondents in this study, some challenges to consider in this step are cost, lack of time and resources, and acceptance from the production managers since the risk is that the production managers do not want to shut down the machines to install sensors. Moreover, Bajic et al. (2021) have investigated several challenges related to the Feasibility test, such as unmatured technology, awareness of what kinds of Industry 4.0 technologies exist, large investments and uncertain financial returns, and lack of strategy.

7.4 Analysis and Technical Development

In this step, the responsibilities for activities are decided to give motivation, which is one enabler for this step (Bengtsson, 2008). But at the same time, according to the empirical findings, defining the responsibility is challenging since smart maintenance technologies are characterized as cross-border technologies involving different departments. Other challenges, according to the empirical findings, are competence, not dedicating people with the right skills, difficulty to choose signals, difficult to use and analyze data, data storage, data visualization, convert data to information that can be used for decision-making, setting limits and seeing what is right and what is outside the tolerance, communication between the sensor on an equipment to the IIoT platform, and to set warning limits to identify deterioration that require component replacement and how long time until that point. According to previous research, some challenges related to this step are poor quality of the collected data, too large datasets to manage, competence about technologies, Cyber Security (Bajic et al., 2021), knowing what kind of data to collect and know what to monitor (Giliyana et al., 2022).

7.5 Action Plan and Implementation

In this step, an action plan should be created to fulfill the activities (Lundgren et al., 2021). Implementation challenges, stated by Badri et al. (2018), are the involvement of end-users and support providers. Furthermore, Bengtsson (2008) has presented several enablers to consider during the implementation step, such as management support, goal setting, communication, using pilot project and keep up motivation.

7.6 Follow-Up and Continuous Improvements

In this step, the result of the implementation needs to be compared to before smart maintenance technologies was implemented, including employee perception and cost. Then, the use of smart maintenance technologies needs to be continuously improved to maintain the benefits. Enablers to consider in this step are, additional training and education might be necessary, additional tools might be needed or steps need to be repeated (Bengtsson, 2008). Related to this step, the empirical findings present several challenges when using smart maintenance technologies, such as ensuring that the right technologies are used in the right way, the risk of replacing broken components based on the output of a technology, without doing root cause analysis, competence in a user stage, and some technologies are quickly accepted while others are slower.

8 Conclusions and Discussion

In this research paper, a conceptual implementation process is proposed, based on empirical findings, the authors' three maintenance research articles, Giliyana et al. (2022), (2023a), and (2023b), as well as maintenance research by other researchers. The empirical findings are based on data from respondents from seven large manufacturing companies at Sweden's forefront of technological development. Two of the case companies are classified as lighthouse factories. Furthermore, challenges and enablers for each process step are presented. This research paper generates new scientific knowledge for academia, regarding implementation, challenges, enablers, as well as benefits of using smart maintenance technologies. As mentioned in Sect. 6, the conceptual implementation process presented in this paper supports the manufacturing industry in what step to take to implement smart maintenance technologies. The limitation of this research paper is that the conceptual implementation process is not tested. The next step is to test the implementation process at one or several manufacturing companies and further improve it. One improvement suggestion is also to improve the process steps by adding additional challenges and enablers for each step.

References

Abidi, M. H., Mohammed, M. K., & Alkhalefah, H. (2022). Predictive maintenance planning for industry 4.0 using machine learning for sustainable manufacturing. *Sustainability, 14*(6), 3387. https://doi.org/10.3390/su14063387

Achouch, M., Dimitrova, M., Ziane, K., Sattarpanah Karganroudi, S., Dhouib, R., Ibrahim, H., & Adda, M. (2022). On predictive maintenance in industry 4.0: Overview, models, and challenges. *Applied Sciences, 12*(16), 8081. https://doi.org/10.3390/app12168081

Ahmad, R., & Kamaruddin, S. (2012a). An overview of time-based and condition-based maintenance industrial application. *Computers & Industrial Engineering, 63*, 135–149. https://doi.org/10.1016/j.cie.2012.02.002

Ahmad, R., & Kamaruddin, S. (2012b). A review of condition-based maintenance decision-making. *European J. Industrial Engineering, 6*, 519–541. https://doi.org/10.1504/EJIE.2012.048854

Ahmer, M., Marklund, P., Gustafsson, M., & Berglund, K. (2022). An implementation framework for condition-based maintenance in a bearing ring grinder. *Procedia CIRP, 107*, 746–751. https://doi.org/10.1016/j.procir.2022.05.056

Alcácer, V., & Cruz-Machado, V. (2019). Scanning the industry 4.0: A literature review on Technologies for Manufacturing Systems. *Engineering Science and Technology, an International Journal, 22*(3), 899–891. https://doi.org/10.1016/j.jestch.2019.01.006

Al-Najjar, B., Algabroun, H., & Jonsson, M. (2018, September 27–30). Smart maintenance model using cyber physical system. In: *International conference on "role of industrial engineering in industry 4.0 paradigm" (ICIEIND)*, , pp.1–6.

Amruthnath, N., & Gupta, T. (2018, April 26–28). A research study on unsupervised machine learning algorithms for early fault detection in predictive maintenance. In: *IEEE 5th international conference on industrial engineering and applications (ICIEA)*, pp. 355–361. https://doi.org/10.13140/RG.2.2.28822.24648

Azevedo, H., Araújo, A., & Bouchonneau, N. (2016). A review of wind turbine bearing condition monitoring: State of the art and challenges. *Renewable and Sustainable Energy Reviews, 56*, 368–379. https://doi.org/10.1016/j.rser.2015.11.032

Badri, A., Boudreau-Trudel, B., & Souissi, A. S. (2018). Occupational health and safety in the industry 4.0 era: A cause for major concern? *Safety Science, 109*, 403–411. https://doi.org/10.1016/j.ssci.2018.06.012

Bajic, B., Rikalovic, A., Suzic, N., & Piuri, V. (2021). Industry 4.0 implementation challenges and opportunities: A managerial perspective. *IEEE Systems Journal, 15*, 546–559. https://doi.org/10.1109/JSYST.2020.3023041

Bengtsson, M. (2008, June). A method for implementing condition based maintenance in industrial settings. In *18th international conference on flexible automation and intelligent manufacturing*.

Bokrantz, J. (2017). *On the transformation of maintenance organisations in digitalised manufacturing*. Licentiate thesis, Chalmers University of Technology.

Bokrantz, J., Skoogh, A., Berlin, C., Wuest, T., & Stahre, J. (2020). Smart maintenance: An empirically grounded conceptualization. *International Journal of Production Economics, 223*, 1–17. https://doi.org/10.1016/j.ijpe.2019.107534

Bona, G. D., Cesarotti, V., Arcese, G., & Gallo, T. (2021). Implementation of industry 4.0 technology: New opportunities and challenges for maintenance strategy. *Procedia Computer Science, 180*, 424–429. https://doi.org/10.1016/j.procs.2021.01.258

Cachada, A., Barbosa, J., Leitño, P., Geraldes, C. A. S., Deusdado, L., Costa, J., Teixeira, C., Teixeira, J., Moreira, A. H. J., Moreira, P. M., & Romero, L. (2018, September 04–07). Maintenance 4.0: Intelligent and predictive maintenance system architecture. In: *IEEE 23rd international conference on emerging technologies and factory automation (ETFA)*, pp.139-146.

Campos, J., Kans, M., & Salonen, A. (2020). A project management methodology for the digitalisation of the industrial maintenance domain. In *Advances in asset management and condition monitoring*. https://doi.org/10.1007/978-3-030-57745-2_52

Carnero, M. C., & Novés, J. L. (2006). Selection of computerised maintenance management system by means of multicriteria methods. *Production Planning & Control, 17*, 335–354. https://doi.org/10.1080/09537280600704085

Chang, Y. S., Choi, H. C., Sung, S. Y., & Mun, S. J. (2016, July 10–14). A study of cloud based maintenance system architecture for warehouse automation equipment. In: *IEEE 5th IIAI international congress on advanced applied informatics (IIAI-AAI)*, . pp.985-990. https://doi.org/10.1109/IIAI-AAI.2016.190

Chong, S., Pan, G.-T., Chin, J., Show, P. L., Yang, T. C. K., & Huang, C.-M. (2018). Integration of 3D printing and industry 4.0 into engineering teaching. *Sustainability, 10*(11), 1–13. https://doi.org/10.3390/su10113960

Erboz, G. (2017). How to define industry 4.0: Main pillars of industry 4.0. In: *Managerial trends in the development of enterprises in globalization era.*

Eswaran, M., Gulivindala, A. K., Inkulu, A. K., & Raju Bahubalendruni, M. V. A. (2023). Augmented reality-based guidance in product assembly and maintenance/repair perspective: A state of the art review on challenges and opportunities. *Expert Systems with Applications, 213*, 118983. https://doi.org/10.1016/j.eswa.2022.118983

Figueiredo, M. J. G., Cardoso, P. J. S., Gonçalves, C. D. F., & Rodrigues, J. M. F. (2014, July 16–18). Augmented reality and holograms for the visualization of mechanical engineering parts. In: *IEEE 18th international conference on information visualisation,* pp. 368-373. https://doi.org/10.1109/IV.2014.17

Flores, M., Maklin, D., Golob, M., Al-Ashaab, A., & Tucci, C. (2018, September 17–19). Awareness towards industry 4.0: Key enablers and applications for internet of things and big data. In: *19th IFIP WG 5.5 working conference on virtual enterprises (PRO-VE).* https://doi.org/10.1007/978-3-319-99127-6_32

Fraser, K., Hvolby, H.-H., & Tseng, B. (2015). Maintenance management models: A study of the published literature to identify empirical evidence a greater practical focus is needed. *International Journal of Quality & Reliability Management, 32*, 635–664.

Frost, T., Nöcker, J., Demetz, J., & Schmidt, M. (2019). The evolution of maintenance 4.0 – what should the companies be focusing on now? In: In IncoME-IV.

Giliyana, S., Salonen, A., & Bengtsson, M. (2022, April 26–29). "Perspectives on smart maintenance technologies – A case study in large manufacturing companies." In: *SPS2022*, pp.255-266. https://doi.org/10.3233/ATDE220145

Giliyana, S., Bengtsson, M. and Salonen, A. (2023a). Perspectives on smart maintenance technologies – A case study in small and medium-sized enterprises (SMEs) within manufacturing industry. In: *16th WCEAM proceedings,* , 5–7 October, 2022, pp.571–581.

Giliyana, S., Karlsson, J., Bengtsson, M., Salonen, A., Adoue, V., & Hedelind, M. (2023b). A Testbed for Smart Maintenance Technologies. In: *7th international congress and workshop on industrial AI and eMaintenance (IAI)*, 13–15 June 2023.

Goodall, P., Sharpe, R., & West, A. (2019). A data-driven simulation to support remanufacturing operations. *Computers in Industry, 105*, 48–60. https://doi.org/10.1016/j.compind.2018.11.001

James, A., Kumar, G., Khan, A., & Asjad, M. (2022). Maintenance 4.0: Implementation challenges and its analysis. *International Journal of Quality & Reliability Management, 40*(7), 1706–1728. https://doi.org/10.1108/IJQRM-04-2021-0097

Kanawaday, A., & Sane, A. (2017, November 24–26). Machine learning for predictive maintenance of industrial machines using IoT sensor data. In: *8th IEEE international conference on software engineering and service science (ICSESS)*, , pp. 87-90. https://doi.org/10.1109/ICSESS.2017.8342870

Kans, M., & Galar, D. (2017, November 28). The impact of maintenance 4.0 and big data analytics within strategic asset management. In *Maintenance, performance, measurement & management* (pp. 96–104).

Kour, H., & Gondhi, N. K. (2020). Machine learning techniques: A survey. In: *ICIDCA2019.*

Labib, A. (2004). A decision analysis model for maintenance policy selection using a CMMS. *Journal of Quality in Maintenance Engineering, 10*, 191–202. https://doi.org/10.1108/13552510410553244

Lee, W. J., Wu, H., Yun, H., Kim, H., Jun, M., & Sutheralnd, J. (2019). Predictive maintenance of machine tool systems using artificial intelligence techniques applied to machine condition data. *Procedia CIRP, 80*, 506–511. https://doi.org/10.1016/j.procir.2018.12.019

Lewis, S. A., & Edwards, T. G. (1997, October 30). Smart sensors and system health management tools for avionics and mechanical systems. In *16th IEEE digital avionics systems conference (DASC)* (pp. 8.5–8.1). https://doi.org/10.1109/DASC.1997.637283

Liu, Y., & Xu, X. (2017). Industry 4.0 and cloud manufacturing: A comparative analysis. *Journal of Manufacturing Science and Engineering, 139*, 1–8. https://doi.org/10.1115/MSEC2016-8726

Lundgren, C., Bokrantz, J., & Skoogh, A. (2021). A strategy development process for smart maintenance implementation. *Journal of Manufacturing Technology Management, 32*(9), 142–166. https://doi.org/10.1108/JMTM-06-2020-0222

Lundgren, C., Bokrantz, J., & Skoogh, A. (2022, April 26–29). Hindering factors in smart maintenance implementation. In: SPS2022, pp.629-637. https://doi.org/10.3233/ATDE220181

Masoni, R., Ferrise, F., Bordegoni, M., Gattullo, M., Uva, A. E., Fiorentino, M., Carrabba, E., & Di Donato, M. (2017). Supporting remote maintenance in industry 4.0 through augmented reality. *Procedia Manufacturing, 11*, 1296–1302. https://doi.org/10.1016/j.promfg.2017.07.257

Masood, T., & Sonntag, P. (2020). Industry 4.0: Adoption challenges and benefits for SMEs. *Computers in Industry, 121*, 103261. https://doi.org/10.1016/j.compind.2020.103261

Matt, D., Modrák, V., & Zsifkovits, H. (2020). *Industry 4.0 for SMEs challenges, opportunities and requirements*. Palgrave Macmillan Cham.

Maxwell, J. (1996). *Qualitative research design: An interactive approach*. Sage Publications, Inc.

Moeuf, A., Lamouri, S., Pellerin, R., Tamayo-Giraldo, S., Tobon-Valencia, E., & Eburdy, R. (2020). Identification of critical success factors, risks and opportunities of industry 4.0 in SMEs. *International Journal of Production Research, 58*, 1384–1400. https://doi.org/10.1080/00207543.2019.1636323

Moubray, J. (1997). *Reliability-centred maintenance: [RCM II]*. Butterworth Heinemann.

Patton, M. (1990). *Qualitative evaluation and research methods*. Sage Publications, Inc.

Prajapati, A., & Ganesan, S. (2013). Application of statistical techniques and neural networks in condition-based maintenance. *Quality and Reliability Engineering International, 29*. https://doi.org/10.1002/qre.1392

Rastegari, A. (2017). *Condition based maintenance in the manufacturing industry: From strategy to Implementation*. Doct. thesis,. Mälardalen University.

Rikalovic, A., Suzic, N., Bajic, B., & Piuri, V. (2021). Industry 4.0 implementation challenges and opportunities: A technological perspective. *IEEE Systems Journal*, 1–14. https://doi.org/10.1109/JSYST.2021.3101673

Roy, R., Stark, R., Tracht, K., Takata, S., & Mori, M. (2016). Continuous maintenance and the future – Foundations and technological challenges. *CIRP Annals, 65*(2), 667–688. https://doi.org/10.1016/j.cirp.2016.06.006

Säfsten, K., & Gustavsson, M. (2020). *Research methodology – for engineers and other problem-solvers*. Studentlitteratur AB.

Silva, D., Carvalho, L. I., Soares, J., & Sofia, R. C. (2021). A performance analysis of internet of things networking protocols: Evaluating MQTT, CoAP, OPC UA. *Applied Sciences, 11*(11), 4879.

Silvestri, L., Forcina, A., Introna, V., Santolamazza, A., & Cesarotti, V. (2020). Maintenance transformation through industry 4.0 technologies: A systematic literature review. *Computers in Industry, 123*, 1–16. https://doi.org/10.1016/j.compind.2020.103335

Singh, S., Galar, D., Baglee, D., & Björling, S.-E. (2013). Self-maintenance techniques: A smart approach towards self-maintenance system. *International Journal of System Assurance Engineering and Management, 5*(1), 75–83. https://doi.org/10.1007/s13198-013-0200-7

Soori, M., Arezoo, B., & Dastres, R. (2023). Internet of things for smart factories in industry 4.0, a review. *Internet of Things and Cyber-Physical Systems, 3*, 192–204. https://doi.org/10.1016/j.iotcps.2023.04.006

Subramaniyan, M., Skoogh, A., Salomonsson, H., Bangalore, P., & Bokrantz, J. (2018). A data-driven algorithm to predict throughput bottlenecks in a production system based on active periods of the machines. *Computers & Industrial Engineering, 125*, 533–544. https://doi.org/10.1016/j.cie.2018.04.024

Tao, F., Qi, Q., Wang, L., & Nee, A. (2019). Digital twins and cyber–physical systems toward smart manufacturing and industry 4.0: Correlation and comparison. *Engineering, 5*(4), 653–661. https://doi.org/10.1016/j.eng.2019.01.014

Witkowski, K. (2017). Internet of things, big data, industry 4.0 – Innovative solutions in logistics and supply chains management. *Procedia Engineering, 182*, 763–769. https://doi.org/10.1016/j.proeng.2017.03.197

Yin, S., & Kaynak, O. (2015). Big data for modern industry: Challenges and trends [point of view]. *Proceedings of the IEEE, 103*(2), 143–146. https://doi.org/10.1109/JPROC.2015.2388958

Yuan, J., & Liu, X. (2013). Semi-supervised learning and condition fusion for fault diagnosis. *Mechanical Systems and Signal Processing, 38*, 615–627. https://doi.org/10.1016/j.ymssp.2013.03.008

San Giliyana is an Industrial Doctoral Student at Mälardalen University and employed at Mälardalen Industrial Technology Center (MITC). He has obtained an MSc in Product and Process Development from Mälardalen University and has worked for many years as a maintenance, production, and logistic lecturer at Mälardalen University and previously as a maintenance engineer at Volvo Group. His research interest is the implementation of smart maintenance technologies in the manufacturing industry.

Antti Salonen is an Associate Professor at Mälardalen University. He obtained his PhD in Strategic maintenance management in 2011 from Mälardalen University. His research interests are dependable production systems, and maintenance as a contributor to sustainable production. One focus area is human factors in discrete manufacturing industry. He has published articles in Journal of Quality in Maintenance Engineering, International Journal of Productivity and Performance Management, Journal of Maintenance Engineering, and International Journal of COMADEM. He has also presented and published 25 conference papers.

Marcus Bengtsson is Associate Professor and has dual employment at Mälardalen University and at Volvo Construction Equipment. He holds PhD (2007) in maintenance management from Mälardalen University. His research interests are maintenance management and maintenance and manufacturing KPIs. His focus is the implementation of various strategies, methodologies and KPIs to increase productivity. He has previously published in the Journal of Quality in Maintenance Engineering, International Journal of Advanced Manufacturing Technology, International Journal of COMADEM and International Journal of Technology and Human Interaction. He has also presented and published papers in more than 40 conferences.

A Framework for Assessing Emerging Technology Risks in Industrial Asset

Issa Diop, Georges Abdul-Nour, and Dragan Komljenovic

Abstract The management of risks in the context of Industry 4.0 is currently lacking accurate and efficient systematic approaches and tools, leading to a potential underestimation or unrealistic perception of risks in various domains where effective risk management is crucial. Traditional methods, while valuable, have limitations and may not adequately capture all the factors that influence system safety. To address the challenges posed by conventional industry issues, emerging risks, and the complexities of socio-technical systems, there is a need for comprehensive Asset Management and Decision Support approaches. These approaches should encompass both conventional and emerging risk safety management, providing innovative and efficient solutions to support practitioners in navigating these complex environments. Based on the rationale provided, this paper is dedicated to the identification and analysis of risk management components, particularly pertaining to emerging safety risks in the context of Industry 4.0. It also examines the challenges posed by extreme, rare, and disruptive events that have the potential to severely impact organizational performance. The research focuses on relatively new methods grounded in system theories, specifically the Functional Resonance Analysis Method (FRAM) and the System-Theoretic Accident Model and Processes (STAMP). These approaches are considered the most suitable for investigating and addressing the research objectives. To validate the efficiency and practicality of the adopted methods, further research initiatives will be focused on conducting case studies. These case studies will aim to gather more accurate data and insights related to the application of FRAM and STAMP in real-world scenarios.

I. Diop (✉) · G. Abdul-Nour
Department of Industrial Engineering, University of Quebec in Trois-Rivieres, Trois-Rivieres, QC, Canada
e-mail: issa.diop.1@etsmtl.net

D. Komljenovic
Hydro-Quebec Research Institute, Varennes, QC, Canada

A. Crespo Márquez et al. (eds.), *Advances in Asset Management: Strategies, Technologies, and Industry Applications*, Engineering Asset Management Review 3, https://doi.org/10.1007/978-3-031-52391-5_4

Keywords Asset management strategy · Enterprise risk management (ERM) · Emerging risks · Extreme-rare-and-disruptive-events · Resilience · Industry 4.0 / 5.0 · Risk-informed decision-making approach (RIDM) · Functional resonance analysis method (FRAM) · System-theoretic accident model and processes (STAMP)

1 Introduction

The rising complexity of socio-technical systems driven by Industry 4.0 presents significant challenges for conventional analysis techniques used to assess safety risks (for e.g., see challenges for Complex System Governance by Keating et al., 2022 as well as Complex system governance as a framework for asset management by Katina et al., 2021). These traditional methods, such as Failure Modes and Effects Analysis (FMEA), Fault Tree Analysis (FTA), Hazard and Operability Analysis (HAZOP), Event Tree Analysis (ETA), and Bowtie analysis, have been valuable in analysing safety risks in various domains. However, as systems become more interconnected, dynamic, and technologically advanced, new approaches are needed to address the emerging risks. Industry 4.0 introduces a range of new technologies, including the Internet of Things (IoT), artificial intelligence, robotics, and cyber-physical systems. These technologies enable automation, data-driven decision-making, and integration across various components of the system. While they bring numerous benefits, they also introduce novel risks that may not be adequately addressed by traditional analysis methods. To effectively manage safety risks in the context of Industry 4.0, new tools and approaches are required. These tools should consider the interdependencies and interactions among various system components, both technological and human, as well as the potential cascading effects and emerging risks associated with complex socio-technical systems.

Developing innovative techniques such as system dynamics modelling, resilience engineering (Hickford et al., 2018; Woods, 2015; Praetorius et al., 2015), systemic risk analysis and managing emerging risks (ISO, 2009, 2018a, b; CEN, 2013), or integrating concepts from Functional Resonance Analysis Method (FRAM) (for e.g., Diop et al., 2022b; Patriarca et al., 2020; Gattola et al., 2018; De Carvalho, 2011), System-Theoretic Accident Model and Processes (STAMP) (for e.g., Allison et al., 2017; Leveson, 2016; Ouyang et al., 2010), and Risk-Informed Decision-Making Approach (RIDM) (for e.g., Gaha et al., 2021; Komljenovic et al., 2019; Dezfuli et al., 2010; Komljenovic et al., 2016; Zio & Pedroni, 2012) can help address the challenges posed by the rising complexity of socio-technical systems driven by Industry 4.0. These approaches emphasize a holistic understanding of the system and its interactions, proactive identification of potential risks, and the ability to adapt and respond effectively to emerging risks. In the same context, Abdul-Nour et al. (2021) have put forth a safety management framework that addresses decision-making in the presence of risk and uncertainty. Their framework, which can be found in **Appendix 1**, offers a systematic approach to enhancing resilience in complex systems.

The primary focus of our researchers is to address the research question that arises from the need to integrate complementary approaches to risk management. With this in mind, the overall objective of this research paper is to develop a Framework for Assessing Emerging Technology Risks in Industrial Assets. This framework aims to capture the unique challenges and uncertainties that arise from the adoption and implementation of new technologies, such as artificial intelligence, Internet of Things, or robotics, within industrial settings.

The remainder of this paper is structured as follows: Sect. 2 summarizes the literature review in asset management complexity and uncertainty as well as the FRAM and STAMP approaches. Section 3 describes the proposed framework and upcoming case-study. Finally, Sect. 4 concludes the study then provides new research directions.

2 Literature Review

2.1 Asset Management Complexity and Uncertainty

Managing assets in today's world is indeed a complex challenge due to various factors. The intense international competition and the unpredictable nature of global markets create a dynamic and uncertain environment for organizations. Additionally, the global landscape is marked by various insecurities, further adding to the complexity of asset management. These factors create a demanding environment where organizations must navigate uncertainties, mitigate risks, and ensure the optimal performance and resilience of their assets. These organizations are constantly confronted with a wide range of risks and uncertainties that have the potential to impact their objectives, as well as the performance of technical and technological systems and human operators. These risks can range from traditional risks to emerging risks that are influenced by various factors such as technological advancements, global changes, and shifting market dynamics. In recent years, there has been an emergence of new types of risks that pose significant challenges to organizations. These risks create conditions that are conducive to the occurrence of extreme, rare, and disruptive events, which can severely disrupt organizational performance. Indeed, the unstable global economic context, the highly insecure political context resulting from the conflict between Russia and Ukraine, and the previous COVID-19 pandemic are compelling asset decision-makers to reassess and modify their economic asset management models. These external factors have created significant uncertainties and challenges for businesses and investors worldwide. The volatile economic conditions, geopolitical tensions, and the far-reaching impacts of the pandemic have necessitated a proactive approach to mitigate risks, identify new opportunities, and adjust strategies accordingly. Asset decision-makers must carefully analyse the evolving landscape and adapt their economic asset management models to navigate these complex and rapidly changing circumstances.

2.2 Functional Resonance Analysis Method

The Functional Resonance Analysis Method (FRAM) is a modern approach to performance assessment used in accident investigation and risk assessment. It aligns with the principles of resilience engineering and embraces the concept of safety II rather than safety I (Hollnagel, 2012, 2014). The safety I approach, which is commonly associated with traditional hazard analysis methods like Failure Mode and Effects Analysis (FMEA) and Hazard and Operability (HAZOP), focuses primarily on identifying potential failures or hazards. These methods typically employ a bottom-up approach to risk analysis (Sun et al., 2022). In contrast, the safety II concept adopted by FRAM emphasizes understanding the essential functions required for the system to achieve its intended purpose. It shifts the focus to the nature of everyday activities rather than solely focusing on failure modes (Hollnagel, 2012). By adopting the FRAM approach, organizations can gain a deeper understanding of how everyday activities and functions contribute to overall system performance and safety, enabling a more comprehensive and proactive approach to risk assessment and management (Diop et al., 2022b). The FRAM structure follows a five-step process, as depicted in Table 1.

Step 0. Purpose of the FRAM analysis.

This step focuses on the objective of the analysis, which can be retrospective, aiming to understand past events such as accident investigation, or prospective, aiming to anticipate future events such as risk assessment.

Step 1. Identification & description of functions.

In this step, the focus is on identifying and describing the essential functions that contribute to the effective functioning of the system. These functions represent the activities being studied and their interdependencies within the system. The couplings among functions reflect the interconnectedness and mutual influence between them. Whether performed by humans, machines, or a combination of both, these functions are crucial for the system to operate effectively and achieve its overall objectives.

In the model, each function is symbolized by a hexagon shape, containing six important aspects: Input (I), Output (O), Preconditions (P), Resources (R), Time (T), and Control (C). These aspects capture the various types of incoming instances that impact the function from other interconnected functions, known as upstream functions. Except for the output, each aspect represents an input that influences the

Table 1 Depiction of the main steps of the FRAM method

Step	*0*	*1*	2	*3*	*4*
Description	Purpose of the FRAM analysis.	Identification and description of functions	Identification of variability	Identification of functional resonance	Determination of safety constraints

(Source: own representation based on Hollnagel, 2012)

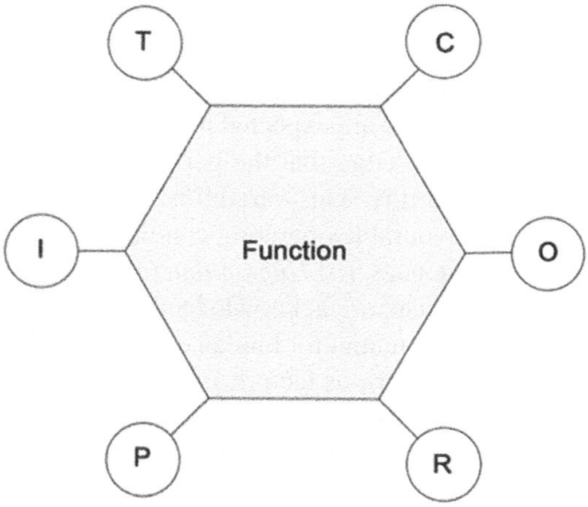

Fig. 1 Illustration of a function in FRAM

function's behavior and operation. On the other hand, the output of a function represents the result or outcome generated by that function, which is then connected to one or more aspects of downstream functions in the system. This interconnectedness illustrates how the functions within the system interact and exchange information to achieve overall system objectives. Figure 1 provides a graphical representation of a function, offering a visual depiction of its structure and components.

Step 2. Identification of variability.

This step forms a crucial basis for further analysis and making informed decisions regarding the management of performance variability. In the original version of the method, it involved assessing a set of Common Performance Conditions (CPCs) to estimate the potential variability. These CPCs included factors such as resource availability, training and experience, communication quality, operational support, availability of procedures, work conditions, conflicting goals, time pressure, circadian rhythm, stress, team collaboration, and organizational support. By evaluating these CPCs, it becomes possible to assess the likelihood of performance variability for each identified function, considering its specific characteristics and nature (Macchi, 2010; Hollnagel, 2004).

To address limitations of the CPCs-based Performance Variability Assessment approach, Hollnagel (2012) introduces an alternative methodology for accurately estimating normal performance variability. This methodology considers the characteristics of different functions and differentiates between two factors that contribute to variability, namely: internal variability and external variability. The model helps determine the extent to which each individual function within the system is influenced by internal (endogenous) or external (exogenous) variability. This approach allows for a more nuanced assessment of performance variability, considering both internal and external factors.

Various function types can be considered, including: *(i) Technological functions*, achieved through different types of technology, are an integral part of the FRAM concept. These functions are typically perceived as stable and predictable, implying that their outputs are expected to remain relatively consistent. However, it is important to acknowledge that the performance of technology-based functions can still exhibit variability. This variability can stem from factors like insufficient maintenance, unfavourable operating conditions, gradual wear and degradation, and other similar influences. *(ii) Human functions*, whether performed individually or collectively by human, are acknowledged as variable and susceptible to instability due to the inherent potential for human error. The variability in human-based functions can be ascribed to various factors, including circadian rhythm, human factors, ergonomics (comprising psychological and physiological aspects), social pressures, decision-making processes, and other pertinent considerations. *(iii) Organizational functions* involve activities performed by groups of individuals within an organization and are described at the organizational level, setting them apart from individual human-based functions. Organizational functions usually exhibit low-frequency variability, suggesting gradual changes over time. However, when variability occurs, it can have a significant impact, often with a considerable magnitude. Multiple factors contribute to the potential variability of organizational functions, including elements such as luck or ineffective communication, conflicting or unclear priorities, inadequate coordination, and other dynamics within the organization. These factors have the capacity to create fluctuations and deviations in the performance of organizational functions. Table 2 and Table 3 presents an overview of the potential variability in the output of functions, specifically in terms of *timing* and *precision* respectively. It provides an outline of how functions within the system may exhibit variations in the *timing* of their outputs and the level of *precision* achieved.

Table 2 Potential output variability pertaining to Time

	Too early	*On time*	*Too late*	*Not at all*
Technological	Unlikely	Normal, expected	Unlikely, but possible if software is involved	Very unlikely (only in case of complete breakdown)
Human	Possible, serendipity	Possible, should be typical	Possible, more likely than too early	Possible, to a lesser degree
Organisational	Unlikely	Likely	Possible	Possible

(Source: adapted from Hollnagel, 2012)

Table 3 Potential output variability pertaining Precision

	Precise	*Acceptable*	*Imprecise*
Technological	Normal, expected	Unlikely	Unlikely
Human	Possible, but unlikely	Typical	Possible, likely
Organisational	Unlikely	Possible	Likely

(Source: adapted from Hollnagel, 2012)

This information can help stakeholders understand and anticipate potential variations in the system's performance, enabling them to make informed decisions and implement appropriate measures to manage and mitigate variability.

Step 3. Identification of functional resonance.

This step is crucial for understanding the interdependencies among functions and their variability within the system. It involves analysing and characterizing the sources of functional resonance by studying the functional aspects. Variability in upstream functions can have a cascading effect, leading to performance variability that propagates through downstream functions. This phenomenon, known as functional resonance. The latter occurs when variability and interactions among functions become intensified, potentially resulting in unexpected behaviours or failures. Exceeding a critical tolerance level of functional resonance intensity can increase the risk of accidents or undesired outcomes within the system. Monitoring and managing functional resonance are crucial for maintaining system safety and preventing adverse outcomes. Through the examination of functional aspects such as input, output, preconditions, resources, time, and control, the analysis allows for the identification of sources that can contribute to functional resonance. This analysis helps determine how variations and interactions among functions can impact the overall performance of the system.

Step 4. Determination of safety constraints.

In this step, the focus is on adjusting performance variability and promoting positive outcomes, rather than solely addressing negative outcomes. Aligning with the four key principles of the FRAM concept, various solutions are recommended. These solutions can range from eliminating risks at their source to implementing safeguards, redundancy, training, feedback mechanisms, or other control measures to mitigate the negative effects of variability and enhance the positive effects. Additionally, strategies may involve changes in processes, procedures, training, technology, or organizational factors to effectively manage variability within the system. These measures aim to improve the system's resilience to adapt to changing conditions and ensure its reliable and safe operation.

In summary, although the FRAM model is valuable for identifying safety constraints, it is essential to recognize that this aspect extends beyond the core principles of FRAM. While the FRAM process provides a macro analysis of the system and offers insights into its overall functioning and dynamics, addressing specific safety constraints requires the incorporation of other methods. In this study, we utilize the STAMP method, as recommended by experts such as Hollnagel (2018) and Leveson (2016). The STAMP approach complements the FRAM model by providing a deeper understanding of the system's control structure and the interactions between components, enabling a comprehensive assessment of safety and risk. By combining the strengths of both FRAM and STAMP, we can effectively address safety constraints and gain a deeper understanding of the system's safety and performance (Diop et al., 2022a, b).

2.3 System-Theoretic Accident Model and Processes

Leveson (2016) introduces a novel approach to accident causation and safety analysis called the System-Theoretic Accident Model and Processes (STAMP). STAMP is a top-down system engineering approach that offers a unique perspective on safety and security analysis. Its theoretical foundation draws upon overall systems theory, allowing for a more comprehensive assessment of highly complex systems compared to traditional safety analysis methods. Traditional accident investigation methods often focus solely on technical and technological aspects, while STAMP takes a broader system thinking approach by considering a comprehensive set of factors and understanding the system as a whole and its dynamic behaviour.

One key distinction of STAMP is its perspective on system safety and security as a *"dynamic control problem"* (rather than a failure or reliability problems). It is important to note that the STAMP process acknowledges that accidents resulting from independent component failures are contained within the system model. In other words, it recognizes that individual component failures alone may not lead to accidents if the safety control system effectively handles those failures and prevents them from propagating or causing harm. However, the focus of the STAMP process goes beyond isolated component failures. It emphasizes the need to identify and address the interactions among components, as well as the control mechanisms that govern those interactions. Accidents can arise when these interactions are not properly managed, leading to the violation of safety constraints and the potential for system-wide failures or hazardous situations. In other words, within the STAMP process, accidents occur when the safety control system fails to effectively manage defective interactions among system components. These defective interactions refer to situations where safety constraints or requirements are violated, leading to unsafe conditions or hazardous events.

STAMP considers not only technical elements but also human operators and organizational aspects. By integrating human and organizational considerations alongside technical and technological aspects into the analysis, STAMP recognizes the crucial role of human operators in system behaviour. It acknowledges that human actions and decision-making can significantly influence the overall system's safety and performance. Additionally, STAMP emphasizes the significance of organizational factors. This includes aspects such as management practices, communication protocols, training procedures, and organizational culture, all of which play a vital role in shaping the system's behaviour and safety outcomes. Indeed, STAMP recognizes that accidents can arise not only from technical failures but also from issues such as inadequate training, ineffective communication, flawed organizational processes, and inadequate safety culture. STAMP recognizes that accidents and failures in complex systems often arise from the dynamic interactions among components, rather than isolated failures of individual components.

The STAMP causality model incorporates a top-down hazard assessment technique known as System-Theoretic Process Analysis (STPA). The latter is an innovative method for analysing hazards based on the extended model of accident causation in STAMP. The primary goal of STPA is to identify accident scenarios that cover the

Table 4 Steps of STPA

Step	*Description*
(1) Defining the purpose of the analysis	Define the purpose of the analysis. Identify the system's boundaries. Identify the system's hazards and losses. Establish safety constraints.
(2) Modelling the control structure (HSCS)	Systems are Hierarchical Safety Control Structures (HSCS) with feedback control loops at different levels, governing system activities and behaviours. Identify the controllers, the controlled Process, the control actions (CA) and the feedbacks.
(3) identifying unsafe control actions (UCAs)	Identifying UCAs involves recognizing behaviours or actions that must be prevented to mitigate system hazards and prevent losses within the sociotechnical system.
(4) identifying loss scenarios	In the final step of the analysis, the emphasis is on comprehending the occurrence and propagation of each UCA identified in step 3, leading to losses within the system. What could cause UCAs? Why would CA be not executed or not followed properly?

entire accident process, going beyond just the electromechanical components involved (Leveson, 2016). The STPA method enables the control of both the system components and the system as a whole, ensuring that safety requirements and constraints are effectively implemented in the operational system. It considers the interactions and behaviours of various system elements to identify potential hazards and evaluate the effectiveness of safety measures. The STPA process consists of several steps, which are illustrated in Table 4. These steps provide a systematic approach to analysing hazards and identifying safety requirements at different levels of the system. By following the STPA process, organizations can gain a comprehensive understanding of potential accident scenarios and develop strategies to mitigate risks and enhance system safety.

(Step 1) Define the Purpose of the Analysis.

The first stage of the analysis process begins by outlining the purpose of the analysis, encompassing the identification of safety requirements and goals set by stakeholders. Within this phase, the system's boundaries are defined, and the hazards and losses that necessitate examination in the study are identified. Safety constraints are subsequently established, considering the system's hazards and losses, thereby establishing the desired safety level for the system. Table 5 presents an example of such losses and hazards as well as high-level safety constraints associated with an aircraft system.

(Step 2) Model the Control Structure.

Systems theory regards systems as Hierarchical Safety Control Structures (HSCS), consisting of feedback control loops functioning at different levels and encompassing various system activities and behaviours. Figure 2 visually presents the dynamic nature of a feedback control loop, emphasizing the interplay between the controlled process and the controller. These loops involve a two-way exchange of information. HSCS sees systems as feedback control loops operating at different levels.

Table 5 Example of an aircraft losses and hazards

Losses Hazards Constraints	*ID*	*Descriptions*
Losses (Ai)	A1	Death or serious injury to aircraft passengers or people in the area of the aircraft
	A2	Unacceptable damage to the aircraft or objects outside the aircraft
	A3	Financial losses resulting from delayed operations
	A4	Reduced sales due to damage to aircraft or airline reputation
Hazards (Hi)	H1	Insufficient thrust to maintain controlled flight [A1, A2]
	H2	Loss of airframe integrity [A1, A2]
	H3	Violating minimum separation between aircraft and fixed or moving objects [A1, A2]
	H4	An aircraft on the ground comes too close to moving or stationary objects or inadvertently leaves the taxiway [A1, A2]
	H5	Aircraft unable to take off when scheduled [A3, A4]
	H6	etc.
Safety constraints (SCi)	SC1	Sufficient thrust must be available to maintain controlled flight [H1]
	SC2	Airframe integrity must be maintained under worst case conditions [H2]
	SC3	Aircraft must satisfy minimum separation standards from fixed or moving objects [H3]
	SC4	Aircraft on the ground must always maintain a safe distance from moving or stationary and objects and remain within safe regions such as taxiways.
	SC5	Aircraft must be able to take off within TBD minutes of scheduled departure [H6]
	SC6	etc.

(Adapted from Leveson & Thomas, 2018)

Downward arrows (Downward Control Channels) enforce safety constraints, while upward arrows (Upward Feedback Channels) provide valuable feedback to inform adjustments. The top level is the controller, responsible for decision-making, and the bottom level is the controlled process regulated by the controller. Figure 3 depicts a basic HSCS model comprising three interconnected control loops that represent control relationships and feedback mechanisms within the system.

HSCS, also referred to as the STPA model, helps address challenges in safety engineering, such as human-system interactions and software behaviour. The *"Control Algorithm"* is the decision-making mechanism used by the controller, based on the *"Process Model (PM)"* or *"Mental Model (MM)"* for human operators. PM influence decision-making and Control Actions (CA). Feedback channels update and refine the PM. The *"Control Algorithm"* may also be referred to as *"Operating Procedures"* or *"Decision-Making Rules"*, providing structured guidelines for the controller's actions. Figure 4 exemplifies a high-level HSCS for *Aircraft system.*

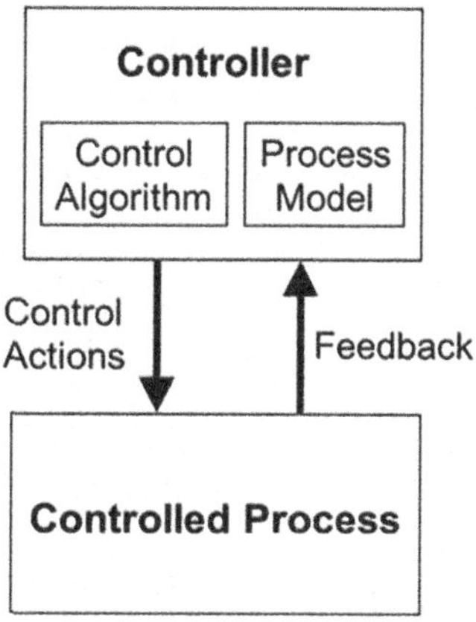

Fig. 2 Representation of a control loop

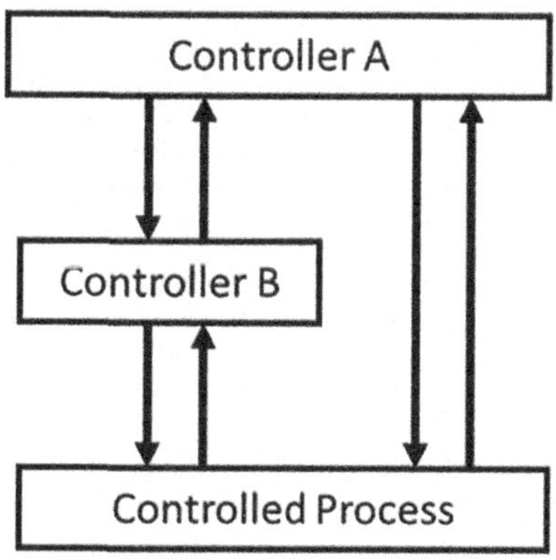

Fig. 3 Representation of a basic HSCS

(Step 3) Identify Potential Unsafe Control Actions (UCAs).

During this analysis phase, the primary focus is on identifying potential Uncontrolled Control Actions (UCAs). UCAs encompass behaviours or actions that need to be prevented to mitigate system hazards and prevent losses within the sociotechnical system. Losses occur when safety constraints are not adequately enforced, implemented, or when these constraints are entirely absent. An example of such losses occurs when the PM does not align with the state of the Controlled Process, resulting in the Controller issuing unsafe commands. Accidents or losses fundamentally arise when the HSCS model fails to effectively manage interactions among system components, leading to a violation of safety constraints. In such instances, the system's ability to ensure the proper functioning and adherence to safety measures is compromised, leading to increased risks and potential negative outcomes. Hence, this step involves a meticulous examination of each Control Actions (CA) identified in the previous step, determining the corresponding UCAs that could lead to undesirable outcomes. By identifying and understanding these potential UCAs,

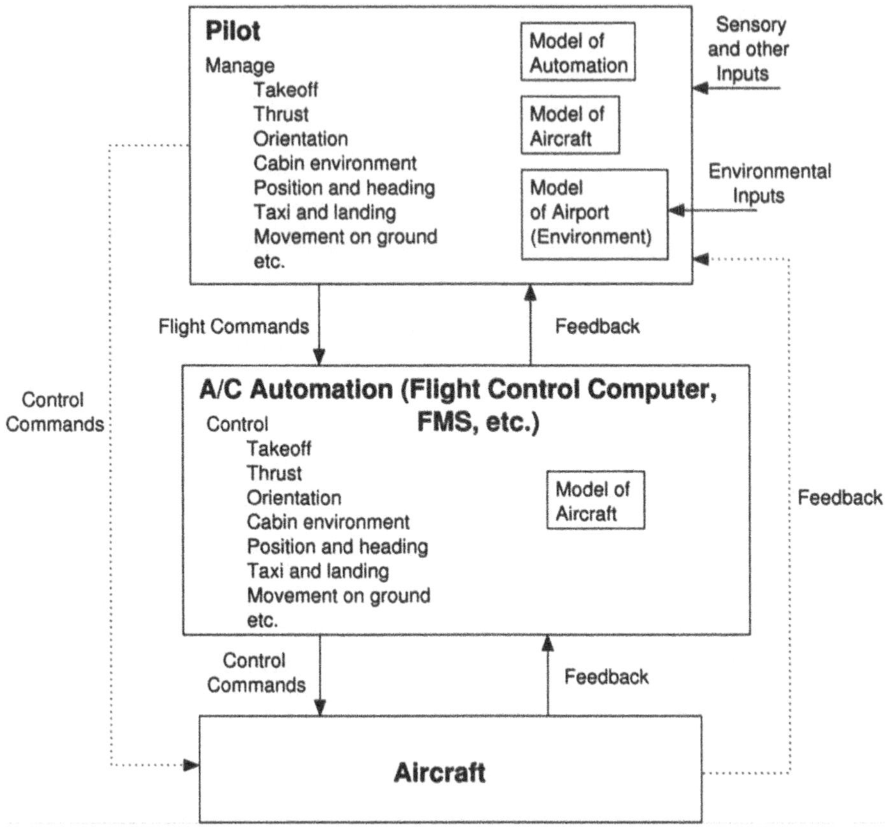

Fig. 4 A High-Level HSCS at the Aircraft Level. (Source: Leveson & Thomas, 2018)

strategies and measures can be developed to prevent or mitigate them, thereby enhancing overall system safety and reliability. Control is not limited to engineering systems or direct human intervention; it also involves policies, procedures, shared values, and organizational culture. These factors shape how the system operates and decisions are made, affecting overall safety and performance. Table 6 outlines four types of UCAs (Leveson, 2016).

Identifying and categorizing these UCAs helps stakeholders understand how CA contribute to system hazards. This understanding aids in developing strategies to mitigate these hazards and improve system safety. Table 7 shows examples of UCAs for an *Aircraft Flight Crew* related to the *Wheel Braking System*. **Appendix 2** depicts the HSCS for the Wheel Braking System.

(Step 4) Identify Loss Scenarios.

After identifying UCAs, the subsequent step is to identify scenarios that could result in losses. A loss scenario encompasses the causal factors that can give rise to UCAs

Table 6 Four distinct types of UCAs

Type of UCAs	*Description*
Not providing causes hazard	This UCA occurs when required CA are not provided, leading to hazardous conditions in the system.
Providing causes hazard	This UCA occurs when executing CA directly leads to hazardous conditions.
Too soon, too late, out of order	This UCA relates to CA being executed too early, too late, or out of order, which can introduce risks and adverse outcomes within the system.
Stopped too soon, applied too long	This UCA pertains to premature or extended duration of CA, which can introduce risks and adverse outcomes within the system.

Table 7 UCA for the Aircraft Flight Crew – (Partial example)

Control Action (CA) by Flight Crew	Not providing causes hazard	Providing causes hazard	Too soon, too late, out of order	*Stopped too soon, applied too long*
Manual braking via brake pedals.	Crew does not provide manual braking during landing, RTO, or taxiing when Autobrake is not providing braking or is providing insufficient braking.	Crew provides manual braking with insufficient pedal pressure. Crew provides manual braking with excessive pedal pressure (resulting in loss of control, passenger/crew injury, brake overheating, brake fade or tire burst during landing). Crew provides manual braking provided during normal takeoff	Crew provides manual braking before touchdown (causes wheel lockup, loss of control, tire burst). Crew provides manual braking too late (TBD) to avoid collision or conflict with another object (can overload braking capability given aircraft weight, speed, distance to object (conflict), and tarmac conditions).	Crew stops providing manual braking command before safe taxi speed (TBD) is reached. Crew provides manual braking too long (resulting in stopped aircraft on runway or active taxiway).

(Adapted from Leveson & Thomas, 2018)

and hazards. The focus here is on understanding (i) "*Why would UCAs occur*"? and, (ii) "*Why would CAs be improperly executed or not executed, leading to hazards*"? This involves examining the factors that contribute to the occurrence of UCAs and the reasons behind the failure or deviation from intended CA. Note that the ultimate aim of STPA is to identify causal scenarios that can result in a hazardous state. Figure 5 presents a control model to assist in causal scenario generation related to a control loop. This visual representation serves as a guide, facilitating further analysis and comprehension of the potential causes and consequences of these flaws within the system.

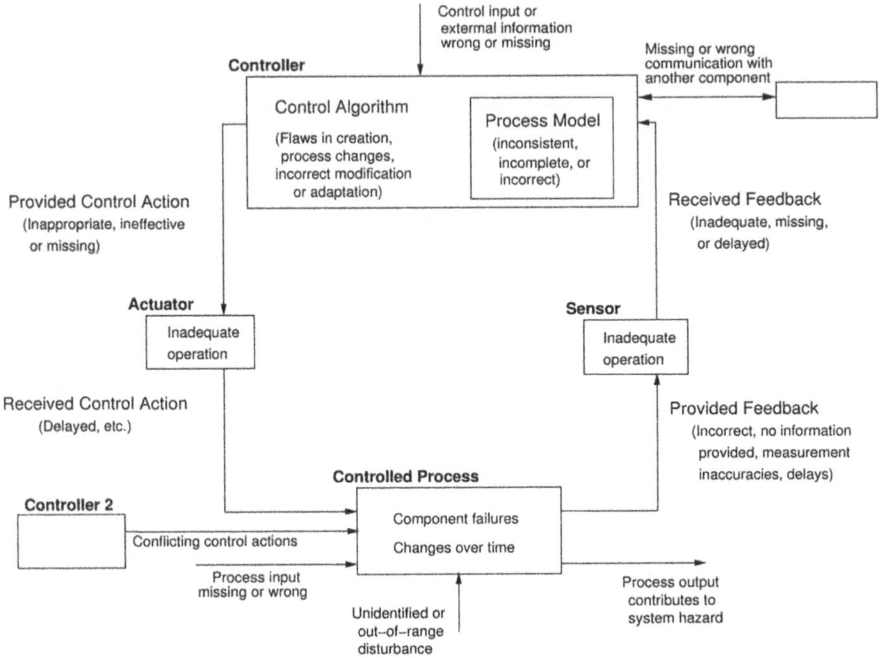

Fig. 5 Control model to support in creating causal scenarios. (Source: Leveson & Thomas, 2018)

3 The Proposed Framework and Upcoming Case-Study

The proposed risk management framework is twofold. It combines the Functional Resonance Analysis Method (FRAM) and the System-Theoretic Accident Model and Processes (STAMP).

Firstly, through the utilization of the FRAM process, we construct a model that emphasizes the variability of functions within the system. This enables us to analyse and comprehend the interdependencies and interactions that exist within the system.

Secondly, we conduct a more in-depth safety assessment by using the STAMP process. We develop a model that governs the behaviour of both the individual components and the system as a whole. This approach empowers us to effectively control and manage the system, ensuring that safety requirements and constraints are upheld throughout its operation. FRAM and STAMP integration provides a comprehensive approach to understanding system functions, variability, and control in relation to potential risks and accidents. It enables organizations to proactively identify and manage risks, enhancing system resilience and safety.

In fine, for upcoming perspectives, we suggest integrating the FRAM and STAMP frameworks into the Risk-Informed Decision Making (RIDM) model. This integrated model provides a comprehensive approach to risk management and decision making, leveraging the strengths of both frameworks. By adopting the RIDM model, organizations can improve their understanding of system behaviour, variability, and control, enabling more informed and effective decision making in managing risks and ensuring system safety and performance. The incorporation of the RIDM model would provide support for long-term performance and the sustainability of an organization, particularly in a dynamic and unpredictable environment. It enables the consideration of risks associated with extreme and rare events within the broader Asset Management strategy and decision-making process. By integrating the RIDM approach, the organization can proactively enhance their ability to assess and address these risks, ultimately contributing to more robust and informed decision-making in asset management. For readers who are unfamiliar with the RIDM approach, we recommend referring to the works of Diop et al. (2021, 2022a, 2023) and their corresponding bibliographic references. These sources provide further details and insights into the RIDM approach, offering a comprehensive understanding of its principles and application.

Figure 6 illustrates the categorization of safety analysis methods, including FRAM, STAMP, and RIDM. These methods are positioned in quadrant 2, which represents highly complex and challenging-to-control systems. This positioning indicates that they are particularly suitable for addressing the unique challenges associated with such systems, providing valuable insights and guidance for managing safety risks.

The LineDrone, showcased in Fig. 7, is one of Hydro-Quebec's Unmanned Aerial Vehicles (UAVs) that will be the focus of our upcoming case study. This state-of-the-art UAV has been specifically designed and optimized for performing direct-contact inspections of high-voltage transmission lines. It offers advanced capabilities and features tailored to meet the specific requirements of this task.

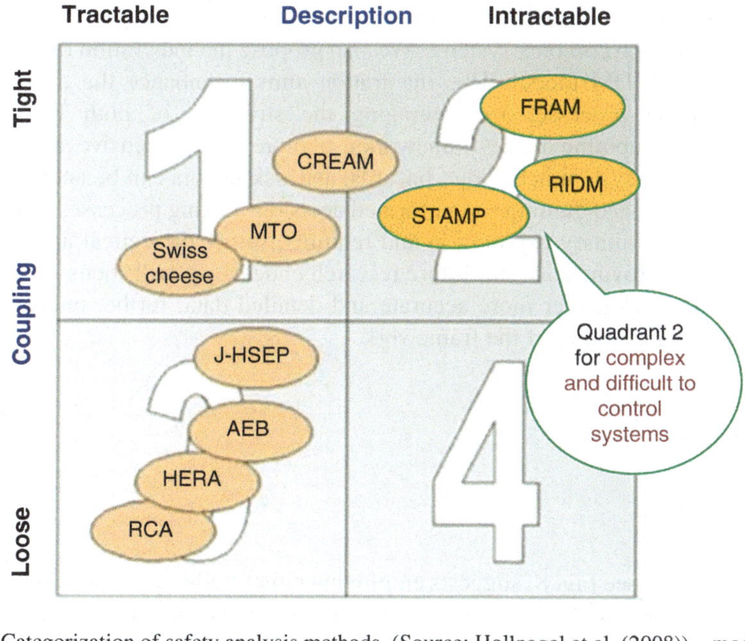

Fig. 6 Categorization of safety analysis methods. (Source: Hollnagel et al. (2008)) – modified)

Fig. 7 Hydro-Quebec Linedrone. (Source: hydroquebec.com)

4 Conclusion

This research aims to present a comprehensive framework for assessing emerging technology risks in industrial assets. To achieve this, we have chosen to utilize a combination of two concepts that we believe are highly effective and valuable in addressing the complexities of socio-technical systems: the FRAM and the STAMP approaches. These methods offer superior capabilities over traditional approaches in engineering such complex systems. We also propose the integration of the framework into the RIDM model. This integration aims to enhance the analysis and decision-making processes by leveraging the strengths of both FRAM and STAMP. By combining these frameworks, a more comprehensive and holistic understanding of system behaviour, hazards, and risk factors can be achieved, ultimately informing and guiding risk-informed decision-making processes. This paper represents a preliminary exploration and requires further theoretical and practical developments. Moving forward, future research endeavours will focus on conducting case studies to gather more accurate and detailed data, further enhancing the applicability and validity of the framework.

Appendices

Appendix 1

This framework (see Fig. 8) suggests employing either traditional risk management, management under uncertainty, or resilience management approaches that acknowledge the complexity of assets (Abdul-Nour et al., 2021). The choice of approach

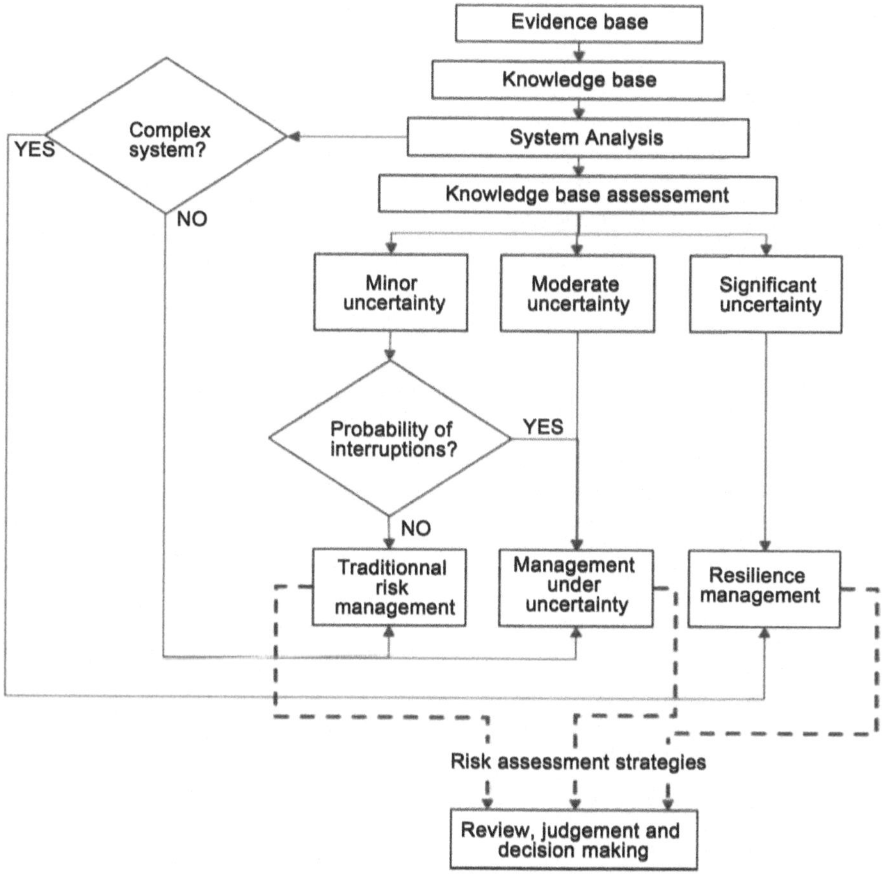

Fig. 8 Risk management of approach. (Source: Abdul-Nour et al., 2021)

will depend on the specific characteristics and context of the assets being managed. *(i) Traditional risk management* involves identifying and assessing risks, implementing controls and mitigation measures, and monitoring and managing risks within predetermined risk tolerances. This approach relies on historical data, probability calculations, and established risk management frameworks. *(ii) Management under uncertainty* recognizes that there are inherent uncertainties and limitations in predicting future outcomes. It emphasizes flexibility, adaptability, and the ability to make informed decisions in the face of uncertain and evolving conditions. It may involve scenario planning, sensitivity analysis, and dynamic decision-making processes. *(iii) Resilience management* focuses on building the capacity to absorb and recover from shocks, disruptions, and unforeseen events. It emphasizes robustness, redundancy, and the ability to bounce back and adapt in the face of adversity. It involves identifying critical assets, diversifying resources, and implementing contingency plans.

Appendix 2

Figure 9 provides a detailed representation of the Hierarchical Safety Control Structure (HSCS) specifically for the Wheel Braking System. It showcases the interconnected control loops and feedback mechanisms that govern the operation and safety of the braking system. The top level of the HSCS represents the controller responsible for decision-making and control, while the bottom level represents the controlled process, which includes the wheel braking system components and their interactions.

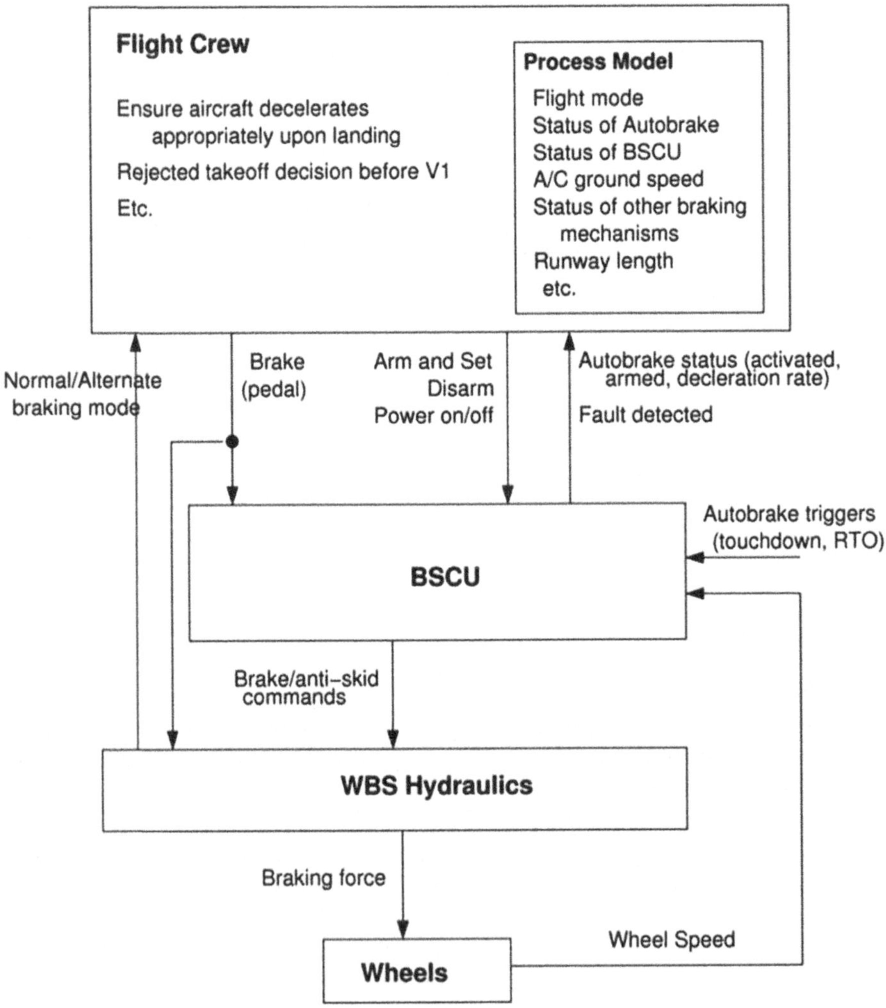

Fig. 9 HSCS for aircraft wheel braking system. (Source: Leveson & Thomas, 2018)

References

Abdul-Nour, G., Gauthier, F., Diallo, I., Komljenovic, D., Vaillancourt, R., & Côté, A. (2021). Development of a resilience management framework adapted to complex asset systems: Hydro-Québec research chair on asset management. In *14th world congress on engineering asset management, WCEAM 2019* (pp. 126–136). Springer Science and Business Media Deutschland GmbH.

Allison, C. K., Revell, K. M., Sears, R., & Stanton, N. A. (2017). Systems theoretic accident model and process (STAMP) safety modelling applied to an aircraft rapid decompression event. *Safety Science, 98*, 159–166.

CEN (European Committee for Standardization). (2013). *Managing emerging technology-related risks* (DIN CWA 16649 (DIN SPEC 91299): 2013-10).

De Carvalho, P. V. R. (2011). The use of functional resonance analysis method (FRAM) in a mid-air collision to understand some characteristics of the air traffic management system resilience. *Reliability Engineering and System Safety, 96*(11), 1482–1498. https://doi.org/10.1016/j.ress.2011.05.009

Dezfuli, H., Stamatelatos, M., Maggio, G., Everett, C., Youngblood, R., & Rutledge, P. (2010). *NASA risk-informed decision-making handbook*. Books Express Publishing.

Diop, I., Abdul-Nour, G., & Komljenovic, D. (2021). Overview of strategic approach to asset management and decision-making. *International Journal of Engineering Research & Technology (IJERT), 10*(12), 64–89.

Diop, I., Abdul-Nour, G., & Komljenovic, D. (2022a). A high-level risk management framework as part of an overall asset management process for the assessment of industry 4.0 and its corollary industry 5.0 related new emerging technological risks in socio-technical systems. *American Journal of Industrial and Business Management, 12*, 1286–1339. https://doi.org/10.4236/ajibm.2022.127071

Diop, I., Abdul-Nour, G., & Komljenovic, D. (2022b). The functional resonance analysis method: A performance appraisal tool for risk assessment and accident investigation in complex and dynamic socio-technical systems. *American Journal of Industrial and Business Management, 12*(2), 195–230.

Diop, I., Abdul-Nour, G., & Komljenovic, D. (2023). Asset and risk management approach in the context of complexity in industry 4.0/5.0 systems. In *16th WCEAM proceedings* (Lecture Notes in Mechanical Engineering.) (pp. 508–520). Springer International Publishing.

Gaha, M., Chabane, B., Komljenovic, D., Côté, A., Hébert, C., Blancke, O., et al. (2021). Global methodology for electrical utilities maintenance assessment based on risk-informed decision making. *Sustainability (Switzerland), 13*(16). https://doi.org/10.3390/su13169091

Gattola, V., Patriarca, R., Tomasi, G., & Tronci, M. (2018). Functional resonance in industrial operations: A case study in a manufacturing plant. *IFAC-PapersOnLine, 51*(11), 927–932.

Hickford, A. J., Blainey, S. P., Hortelano, A. O., & Pant, R. (2018). Resilience engineering: Theory and practice in interdependent infrastructure systems. *Environment Systems and Decisions, 38*(3), 278–291.

Hollnagel, E. (2004). *Barriers and accident prevention*. Ashgate.

Hollnagel, E. (2012). *FRAM: The functional resonance analysis method: Modelling complex socio-technical systems*. Ashgate Publishing Ltd..

Hollnagel, E. (2014). *Safety-I and safety-II. The past and future of safety management*. Ashgate.

Hollnagel, E. (2018). *The functional resonance analysis method*. A brief Guide on how to use the FRAM. https://functionalresonance.com/onewebmedia/Manual%20ds%201.docx.pdf

Hollnagel, E., Pruchnicki, S., Woltjer, R., & Etcher, S. (2008). *Analysis of comair flight 5191 with the functional resonance accident model*. In *8th International symposium of the Australian aviation psychology association* (pp. 8). Sydney, Australia.

ISO (International Organization for Standardization). (2009). *ISO GUIDE 73:2009 risk management—Vocabulary* (pp. 15): Technical Committee: ISO/TMBG Technical Management Board – groups.

ISO (International Organization for Standardization). (2018a). *ISO 31000 risk management—guidelines* (pp. 16). International Organization for Standardization ISO Geneva, Technical Committee: ISO/TC 262 Risk management.

ISO (International Organization for Standardization). (2018b). *ISO 31050 guidance for managing emerging risks to enhance resilience*. International Organization for Standardization ISO Geneva: Work Group 8 (WG8) of the Technical Committee TC262.

Katina, P. F., Pyne, J. C., Keating, C. B., & Komljenovic, D. (2021). Complex system governance as a framework for asset management. *Sustainability, 13*(15), 8502.

Keating, C. B., Katina, P. F., Chesterman, C. W., & Pyne, J. C. (2022). Future challenges for complex system governance research and practice. In *Complex system governance* (pp. 541–575). Springer.

Komljenovic, D., Gaha, M., Abdul-Nour, G., Langheit, C., & Bourgeois, M. (2016). Risks of extreme and rare events in asset management. *Safety Science, 88*, 129–145.

Komljenovic, D., Abdul-Nour, G., & Boudreau, J.-F. (2019). Risk-informed decision-making in asset management as a complex adaptive system of systems. *International Journal of Strategic Engineering Asset Management, 3*(3), 198–238.

Leveson, N. G. (2016). *Engineering a safer world: Systems thinking applied to safety*. The MIT Press.

Leveson, N. G., & Thomas, J. P. (2018). *STPA handbook*. The MIT Press.

Macchi, L. (2010). A Resilience Engineering approach for the evaluation of performance variability: Development and application of the Functional Resonance Analysis Method for air traffic management safety assessment (Doctoral dissertation, École Nationale Supérieure des Mines de Paris).

Ouyang, M., Hong, L., Yu, M.-H., & Fei, Q. (2010). STAMP-based analysis on the railway accident and accident spreading: Taking the China–Jiaoji railway accident for example. *Safety Science, 48*(5), 544–555.

Patriarca, R., Di Gravio, G., Woltjer, R., Costantino, F., Praetorius, G., Ferreira, P., et al. (2020). Framing the FRAM: A literature review on the functional resonance analysis method. *Safety Science, 129*, 104827. https://doi.org/10.1016/j.ssci.2020.104827

Praetorius, G., Hollnagel, E., & Dahlman, J. (2015). Modelling vessel traffic service to understand resilience in everyday operations. *Reliability Engineering & System Safety, 141*, 10–21.

Sun, L., Li, Y.-F., & Zio, E. (2022). Comparison of the HAZOP, FMEA, FRAM, and STPA methods for the Hazard analysis of automatic emergency brake systems. *ASCE-ASME Journal of Risk and Uncertainty in Engineering Systems, Part B: Mechanical Engineering, 8*(3), 031104.

Woods, D. D. (2015). Four concepts for resilience and the implications for the future of resilience engineering. *Reliability Engineering and System Safety, 141*, 5–9. https://doi.org/10.1016/j.ress.2015.03.018

Zio, E., & Pedroni, N. (2012). *Overview of risk-informed decision-making processes*. FonCSI.

Issa Diop is on an impressive academic journey, pursuing a Ph.D. in Industrial Engineering with a focus on asset management and risk assessment at the Université du Québec à Trois-Rivières. He is affiliated with the Hydro-Québec Chair for Asset Management. His research delves into emerging technological risks, showcasing a deep interest in the field. With a Master's in Automated Manufacturing Engineering, Issa has practically integrated lockout-tagout procedures into manufacturing planning. Not stopping there, he's also pursuing a DESS in Resilience, Risks, and Disasters from UQAM. Beyond academics, Issa actively contributes to scientific committees and has numerous publications, earning recognition and awards.

Georges Abdul-Nour is an Emeritus Professor at the Department of Industrial engineering, Université du Québec à Trois-Rivières, Canada. He received his Bachelor and Master Degrees in IE from the Université de Moncton, NB, Canada. He completed his PhD at Texas Tech University; Lubbock Texas, U.S.A. Dr. Abdul-Nour authored and co-authored 10 books and book chapters, and more than 450 refereed papers in international journals and conferences proceeding, and technical reports. His research interest is in fields such as Intelligent Asset Management, Complexes Systems Resilience, Reliability and Maintenance, Intelligent Manufacturing, I4.0 and I5.0, Lean-Six Sigma and JIT (TPS). He is a member of the Editorial Board of the *International Journal of Production Research* (IJPR), and Fellow and board member of ISEAM. Professor Abdul-Nour is a registered member of the Order of Engineers of Quebec (OIQ). He served as Chair of Engineering School and of Research Institute on Small and Medium Enterprises (SME), INRPME at UQTR. He is an elected member of the Academy of Industrial Engineering of Texas Tech. Dr. Abdul-Nour hold the HQ/NSERG Research Chair in Asset Management.

Dragan Komljenovic received a BSc from the University of Tuzla, an MSc from the University of Belgrade, a first PhD from Université Laval in 2002 and a second PhD from Université du Québec à Trois-Rivières in 2018. He works as a researcher at Hydro-Québec's research institute (IREQ) in the field of reliability, resilience, asset management and risk analysis. He has also worked as a reliability and nuclear safety engineer at Gentilly-2 nuclear power plant, Hydro-Québec. Dragan collaborates with several universities and has published more than 100 refereed journal and conference papers. He is a Fellow of the International Society of Engineering Asset Management (ISEAM) and Vice President of the Montreal Chapter of Society of Reliability Engineers (SRE). Dragan has the status of professional engineer in the province of Quebec, Canada.

Part III
Asset Health and Maintenance Strategies

Challenges on an Asset Health Index Calculation

Eduardo Candón Fernández, Adolfo Crespo Márquez, and Antonio Jesús Guillén López

Abstract In the current era of Industry 4.0, we find ourselves in the midst of a profound transformation in the industrial landscape. This new era brings with it a host of challenges and problems, particularly in relation to the effective capture and processing of data. The success of this revolution hinges on our ability to harness data in a meaningful way, but achieving this goal is no small feat.

At the core of this data-driven revolution lies the critical importance of capturing data accurately. However, in many companies, this proves to be an incredibly complex problem. It is not simply a matter of capturing as much data as possible from the moment an asset or system is initiated. Rather, the focus is on acquiring a minimum amount of data that is sufficient to enable proper processing and analysis. This requirement presents a unique challenge in itself, as it often necessitates estimating this minimum data requirement based on a solid and reliable foundation of existing information.

The consequences of lacking adequate information can be far-reaching. Insufficient data availability inevitably leads to deviations in the processing and analysis of the captured data. However, this limitation also offers an opportunity for comparison. By examining assets of the same type that face similar challenges in data capture and processing, valuable insights can be gained. For instance, consider the scenario of comparing the health index of multiple transformers located in different electrical substations and operating under diverse conditions. If the data capture relating to the operational and maintenance variables is equally deficient across these transformers, and similar estimation techniques are employed, it becomes possible to compare the overall health of these equipment units.

To delve deeper into this topic, let us explore the specific example of calculating the Health Index for different pumps. In this particular case, the challenge arises from the fact that the start-up of these pumps predates the availability of operation and maintenance data. Consequently, due to this lack of information, a different

E. Candón Fernández (✉) · A. Crespo Márquez · A. J. Guillén López
Department of Industrial Management, University of Seville, Seville, Spain
e-mail: ecandon@us.es; adolfo@us.es; ajguillen@us.es

A. Crespo Márquez et al. (eds.), *Advances in Asset Management: Strategies, Technologies, and Industry Applications*, Engineering Asset Management Review 3, https://doi.org/10.1007/978-3-031-52391-5_5

approach must be taken. The estimation of various fundamental variables becomes necessary to facilitate the calculation of the Health Index and derive meaningful insights into the condition and performance of the pumps.

In conclusion, the advent of Industry 4.0 has brought forth a range of challenges and problems in the realm of data capture and processing. The ability to obtain and process data accurately is a critical factor in the success of this revolution. However, the complexity of the task lies not only in capturing a substantial amount of data but also in determining the minimum data requirements for meaningful analysis. Despite the difficulties posed by limited information, the comparison of similar assets facing data capture challenges can provide valuable insights. Through a specific example involving pump health index calculations, we can further understand the importance of addressing data estimation and processing in the context of Industry 4.0. Throughout this paper, the example of calculating the Health Index of different pumps will be developed in which the start-up of these goes back to times prior to the date of capture of the operation and maintenance data. Due to this lack of information, it will be necessary to start from the estimation of different fundamental variables for the processing of the data to be calculated.

1 Introduction

An Asset Health Index (AHI) is an asset score, which is designed, in some way, to reflect or characterize the asset's condition and thus, its performance in terms of fulfilling the role established by the organization (De la Fuente et al., 2021). An AHI represents a practical method to quantify the general health of a complex asset. For simple assessments, Condition Based Maintenance (CBM) technologies can precisely estimate the status of a specific asset with defined and specifics failure modes. However, most of these assets are composed of multiple subsystems, and each subsystem can be characterized by multiple modes of degradation and failure. From a pure theoretical perspective, every failure mode of every item that composes a system can be modelled and estimated. In some cases, it may be considered that an asset has reached the end of its useful life, when several subsystems have reached a state of deterioration that prevents the continuity of service required by the business (Hjartarson & Otal, 2006). This calculation can be complex and cause a significant investment in time and resources. It is in the case of complex systems where the health index, based on the results of operational observations, field inspections and laboratory tests, produces a single objective and quantitative indicator. It may be used as a tool to manage assets, to identify capital investment needs and maintenance programs, allowing (Naderian et al., 2008; Naderian et al., 2009; Azmi et al., 2017): (1) Compare the health of equipment located in similar technical locations, to study possible premature deterioration and optimize operation plans and/or asset maintenance if necessary; (2) Communicate more accurately with manufacturers/builders, to understand the behaviour of assets of different manufacturers/builders in specific technical locations; and (3) Support decision-making processes in future investments in assets, or in extension of the life of these (Silvestri et al., 2020). Thus, AHIs are widely used in supporting maintenance and replacement strategies based on asset condition and performance in some countries, to justify asset replacement schemes to the regulators (GB DNO groups, 2017; Australian Local Government Association, 2015; Federation of Canadian Municipalities (FCM) & other seven partner organizations, 2020).

A proper design of a health index should meet the following requirements (Hjartarson & Otal, 2006):

- The index must be indicative of the suitability of the asset to provide continuity to the service and representative of the general health of the asset.
- The index should contain objective and verifiable measures of the condition of the asset, instead of subjective observations.
- The index must be understandable and easily interpreted.

Several methods and models fulfilling these requirements have been reviewed, for instance, the ones by Kinetrics (), DNV GL (Vermeer et al., 2015), Terna (Scatiggio & Pompili, 2013) and GB DNO (GB DNO groups, 2017).

Although most of these models build a streamlined approach to introduce different influent factors to estimate the lifetime expectation/remaining useful life of an asset, several drawbacks are still present in their model formulation:

(i) The AHI procedure seems not to be properly robust from the scientific perspective, as original models are built mostly by practitioners in specific sectors with very specific assets.
(ii) Many influent factors are evaluated based on assumptions that are never discussed (e.g., ranges of numerical values are given as scales for different factors while it is almost unclear what is the basis to define such ranges).
(iii) The procedure proposed is mainly presented in its development and never demonstrated completely with, at least, some case-based reasoning or at the minimum a complete industrial case which would enable a proper validation of the AHI model proposed.

There are approaches in the relevant literature to identify asset health (López et al., 2019) used mainly in CBM applications based on dynamic health assessment, but the concept is different from the one used in this paper, now the health assessment allows comparison and decision-making among different assets.

To overcome these weakness points, in this paper the methodology adopted to model the AHI is only loosely based on the OFGEM Network Asset Indices Methodology (GB DNO groups, 2017) (a similar approach as in the example previously presented in (Crespo et al., 2020)). This method is selected because it is considered simple for simulation model building purposes and very practical in its implementation, if a more robust scientific design of the model format is reached.

More precisely, the method (GB DNO groups, 2017) requires: (1) The identification of the asset, which includes the category of the equipment under study, the current age, the expected life, the name of the manufacturer/builder, the model of the equipment and the location of the installation; (2) The operation and maintenance data recorded during a certain period of time; and (3) The condition of the equipment, that is, the results of the analyses performed on the equipment in site, results of readings of physical variables, results of visual inspections, etc.

The health index model adopted in this paper contains values between 1 and 10, thus being able to compare health between different types of assets. There are other indices that go from 0 to 1 and others that range from 1 to 100. In any case, they all have the same functionality: normalize the health of different assets to be able to compare them with each other.

2 AHI Modelling Methodology

The application procedure for calculating the health index is based on 6 consecutive steps, in which, starting from a design life associated with an equipment's category, a current health index is reached. For this, a series of factors related to the location, operation and condition of the asset are considered. It is presented in the following Fig. 1, the model, with the 6 steps for calculating the health index of an asset. For a precise description of the methodology of the AHI the reader is addressed to (Serra et al., 2019). In addition, for a precise description of the mathematical formulation of the model, the reader is addressed to (Crespo, 2022).

A synthesis of formulation is as follows in Fig. 1.

The methodology for calculating the asset health index consists of 6 steps (Fig. 1) which are briefly described as follows:

Step 1: Selection and Definition

In this initial step, the asset of interest is identified, and its class and subclass are defined. This involves gathering relevant information about the asset, such as its functional location and an estimation of its expected normal life. This step sets the foundation for the subsequent calculations.

Step 2: Evaluation of Load and Location Factors

The next step involves evaluating the impact of load and location factors on the asset's life. Load factors consider the magnitude and frequency of stress or strain that the asset experiences during operation. Location factors take into account environmental conditions or specific operating conditions that may affect the asset's longevity. By assessing these factors, an estimation of the asset's remaining life can be calculated.

Step 3: Calculation of Aging Rate

Assets experience aging phenomena over time, such as corrosion, wear, oxidation, and insulation breakage. The aging rate represents the mathematical expression of this behavior, typically exhibiting an exponential pattern. By determining the asset's aging rate, it becomes possible to quantify the rate at which the asset's condition deteriorates as it ages, considering the various aging-related factors it may encounter throughout its useful life.

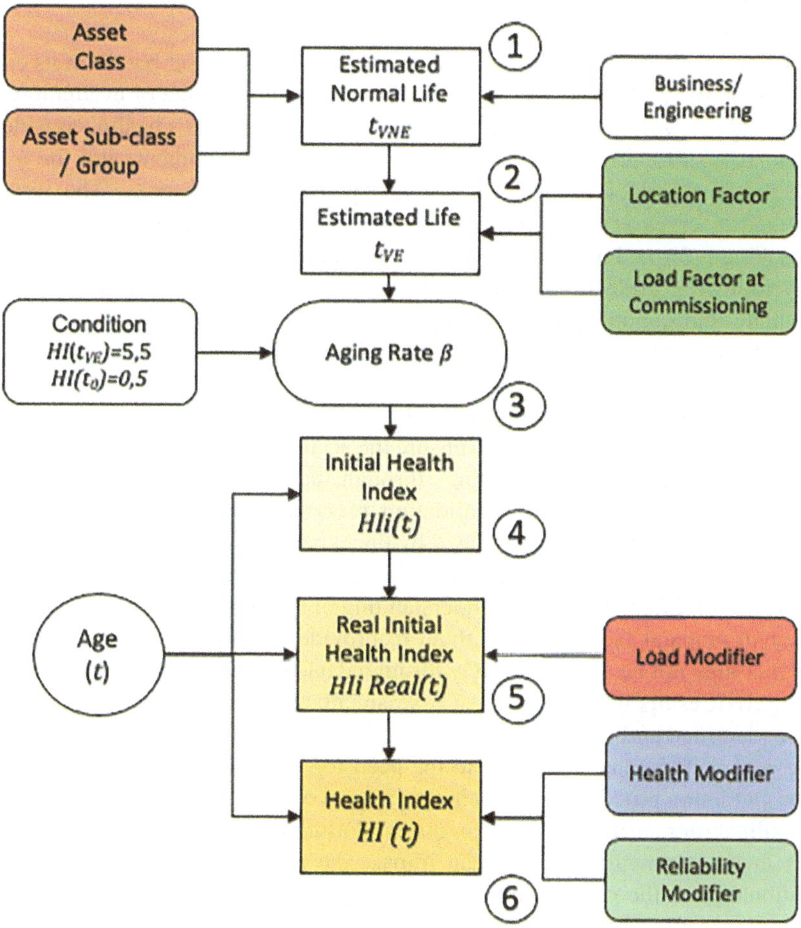

Fig. 1 Procedure to calculate the AHI

Step 4: Obtaining the Initial Health Index

The initial health index is calculated to provide an assessment of the asset's health at its current age. It is a dimensionless number typically ranging between 1 and 10. The initial health index is closely tied to the asset's age, as defined by the aging rate. It reflects the asset's expected condition based on its age and the underlying aging behavior.

Step 5: Obtaining the Actual Initial Health Index

To refine the initial health index further, it is necessary to consider the load modifier recorded for the current age of the asset. This load modifier accounts for any deviations from normal operating conditions that may impact the asset's health. By adjusting the initial health index with this load modifier, the actual initial health index is derived, providing a more accurate representation of the asset's current health status.

Step 6: Obtaining the Health Index

The final step involves determining the health index of the asset based on its condition, operating conditions, and reliability conditions at the time of evaluation. The health index provides a comprehensive assessment of the asset's health, considering factors such as its physical condition, the operating environment, and the asset's reliability in terms of its performance and potential failure risks. The resulting health index serves as a valuable metric for evaluating the overall health of the asset and making informed decisions regarding maintenance or replacement actions.

By following these steps, the calculation of an asset's health index provides a systematic and quantitative approach to assess the asset's condition, estimate its remaining useful life, and guide decision-making regarding maintenance strategies and asset management.

The calculation of the asset health index (AHI) is an important step in assessing the condition of assets. However, to ensure the accuracy and reliability of the results, it is crucial to validate these findings through the expertise of the organization's knowledgeable personnel. This validation process involves comparing the AHI results with the health status estimated by the experts.

The expertise of the organization's personnel is invaluable in assessing the assets' health and condition. Their deep understanding of the assets, operational processes, and environmental factors allows them to provide valuable insights and observations that may not be captured solely through data-driven calculations. By comparing the AHI results with the experts' assessment, any discrepancies or mismatches can be identified and analyzed.

Several factors may contribute to the need for recalibration of the model based on the validation process. Changes in operational or environmental conditions, such as modifications in maintenance practices, variations in operating parameters, or shifts in environmental factors, can impact the asset's health and performance. Additionally, as the organization's knowledge and experience about these assets increase, it may lead to new insights and understanding that require adjustments to the AHI model.

Recalibrating the model based on the validation results and updated knowledge ensures that the AHI remains accurate and reflects the true health state of the assets. This iterative process of validation, comparison, and recalibration helps refine the model over time, enhancing its effectiveness and reliability in assessing asset health.

In summary, the validation of AHI results through expert assessment is essential for ensuring the accuracy and reliability of the calculated health index. By comparing the AHI with the experts' knowledge and experience, discrepancies can be identified and addressed. Recalibrating the model when necessary, in response to changes in operational or environmental conditions or the accumulation of new knowledge, helps maintain the accuracy and relevance of the AHI model, ultimately supporting effective asset management decision-making.

3 Case of Study

The motors of two motor pumps of a power generation plant have been selected as real examples on which the proposed methodology has been applied. The calculation of the health index will make it possible to know the current condition of the assets, making it possible to compare them objectively with each other. This indicator will make it possible to prioritise interventions, care and/or the renewal of the assets analysed.

3.1 Application of the Methodology Proposed

As is shown in Fig. 1, in step 1 a design theoretical life for every asset depending on the equipment category is defined. Its design life can be adapted by the owner according to accumulated experience and the information provided by different manufacturers and builders. In this case of study, it has been considered that it can operate 24 hours a day, every day of the year and for 10 years, so that an estimated normal life of 87,600 hours is left.

In step 2, the estimated owner life can then be adjusted according to the characteristics of the asset location and loading. In the installation where the analysis is carried out, the assets are located indoors, which means that distance to the coast, altitude above sea level, annual average of outside temperature, exposure to corrosive atmosphere or exposure to dust in suspension are factors that do not almost affect the deterioration of the equipment. Therefore, the location factor (F_E) is considered to have no influence, i.e. it is equal to 1. The load factor (F_{EL}) measures the load request that is made on the asset in that location, in front of the maximum admissible load. In this case, the variable selected to calculate the load factor is the flow rate. The values for nominal and maximum allowable flow rate are available in the pump operating manual, resulting in a load factor of 81%. Eq. ((1)) shows the calculation of the estimated life of these pumps:

$$Estimated\ life = t_{EL} = \frac{t_{DL}}{F_{FL}\ x\ F_{EL}} = \frac{87{,}600}{1\,x\,0.81} = 108{,}148\ hours \quad (1)$$

A fundamental hypothesis of the methodology is that the irreversible degradation of an asset follows an exponential behaviour with respect to its age, and in step number 3, the aging rate (β) of the asset is determined by the natural logarithm of the quotient between the asset health index when new (H_{new}) and the asset health index when reaching its expected life ($H_{estimated\ life}$). The equation for its calculation is the following, used in step 3:

$$\beta = \frac{\ln \frac{HI_{new}}{HI_{estimated\ life}}}{Estimated\ life} = \frac{\ln \frac{0,5}{5,5}}{t_{EL}} \tag{2}$$

Then, in step 4, the initial health index (HIi_t) is considered as a dimensionless number between 1 and 10, with an exponential behaviour with respect to the age "t" of the asset, which is characterized by the aging rate as follows:

$$HIi_t = HI_{new} \bullet e^{\beta \bullet t} \tag{3}$$

The health index (HI) is the result of adjusting the initial health index, using load, health, and reliability modifiers. In a first step, the initial health index (HIi_t) of an asset is modified to obtain what we call the real initial health index (HI_iReal) in Step 5, considering the load modifier registered for the current age ($M_L(t)$), using the following equation:

$$HI_i Real(t) = HI_{new} \bullet e^{\frac{\beta \bullet t}{M_L(t)}} \tag{4}$$

where the load modifier is the quotient between the load factor existing at an instant ($F_{RL}(t)$) and the expected load factor (F_{EL});

$$M_L(t) = F_{EL} / F_{RL}(t) \tag{5}$$

The load modifier is a health modifier of the asset, which is considered in this initial phase since it is very likely that in many assets the load recorded during each asset age will be significantly different to the one initially planned for the functional location. The introduction of HIiReal then allows the current asset degradation to be adjusted to compare with the anticipated degradation for the functional location.

Finally, in step 6, the health index of the asset is determined by its operating conditions and reliability conditions at the time of the evaluation. To determine the health index, the following equation is used:

$$HI(t) = HI_i Real(t)^{e^{(MH(t)+MR(t))}} \tag{6}$$

Where:

MH(t): is the asset's health modifier (condition and operation).
MR(t): is the asset's reliability modifier.

For the evaluation of the health modifier (MH) that appears in this last equation, the different variables that can be measured and quantified for each asset sub-category are considered, and that, being independent in their impact on health, add information about it. From the large number of variables available in the plant information system, it is necessary to perform data mining to determine which variables to select as pump health modifiers. To do this, RapidMiner Studio software was used to preprocess the available database, eliminating missing data and outliers, creating a single database, and to analyse the correlation between variables, thus allowing the most representative variables to be selected to determine the final health modifier. In this case, the health modifiers are composed of the operating parameters of speed, flow rate, suction pressure, discharge pressure and suction temperature. These variables are obtained in real time from the PI System.

For the reliability modifier, depending on the sub-category of asset, model and manufacturer, tables can be prepared with the value of this parameter. In this case, the reliability modifiers are made up of the unavailability of the pump and the number of major maintenance or overhauls that are carried out.

Once the proxy variables for health and reliability modifiers have been determined, the challenge is to determine how they impact on the health of the asset. To do this, it will be necessary to determine, within a range of [1; 1.4], how each of these variables affects the health of the pump in a particular way, considering a value of 1 as having no effect on health and 1.4 as having a 40% negative effect on the health of the pump. Likewise, once these ranges have been determined for the different operating thresholds of each variable, they are dimensioned and the modifiers $MH_j(t)$ and $MR_k(t)$ are calculated, respectively, in a range [0;1], which multiplied by the weights of each modifier will give rise to the variable modifier (see Eq. (7) and Eq. (8)). In order to determine the effect of the modifiers, the participation of the organisation's expert group is necessary, which, being familiar with the assets analysed, makes it possible to quantify how each variable affects the asset health. In this case, operational thresholds are established for each variable, and the corresponding modifiers are determined. Table 1 shows the results proposed for each variable.

In addition, Table 2 shows the weights of each of the variables and the associated coefficient γ_i.

The equations to obtain the value of the health modifier (MH) and the reliability modifier (MR) will be the following:

$$MH(t) = \sum_{j=n}^{j=1} \gamma_j \bullet MH_j(t) \tag{7}$$

With:

j = 1... *n* is an index used for different health modifiers.
γ_j: is the weight assigned to the health modifier j in the model.
MH_j (t): is the health modifier at time t, age of the asset.

And

$$MR(t) = \sum_{k=m}^{k=1} \gamma_k \bullet MR_k(t) \tag{8}$$

Table 1 Health and reliability modifiers

Pump variables	Below the admissible range	In the recommended range		Above the admissible range	
Flow	0	0		1	
Suction Pressure	1	0		0	
Discharge Pressure	0	0		1	
Suction Temperature	1	0		1	
	Below the admissible range during t > = 30'	**Below the admissible range during t < 30'**	**In the recommended range**	**Above the admissible range during t < 30'**	**Above the admissible range during t > = 30'**
BFP Speed	0.5	0.25	0	0.5	1
	From 0% to 50%	**From 50% to 75%**		**From 75% to 100%**	
Inactivity	0	0.5		1	
	From 0 to 3	**From 3 to 5**		**More than 5**	
Overhauls	0	0.33		1	

Table 2 Relative weight and coefficient γi

Modifier	Relative weight	Coefficient γ_i
Flow	15.22%	0.046
Suction Pressure	15.22 %	0.046
Discharge Pressure	14.13 %	0.042
Suction Temperature	13.04 %	0.039
BFP Speed	15.22%	0.046
Inactivity	13.04%	0.039
Overhauls	14.13%	0.042

With

k = 1… *m* is an index used for different reliability modifiers.
γ_k: is the weight assigned to the reliability modifier k in the model.
MR_k (t): is the reliability modifier at time t, age of the team.

The determination of the weight assigned to each modifier must be done relative to the rest of the modifiers and assuming a maximum possible impact of the set of modifiers in their worst condition. This is achieved considering the following restrictions:

The sum of the totality of the modifier weights will be equal to the so-called maximum impact rate γ (for this rate, all modifiers always take the value 1).

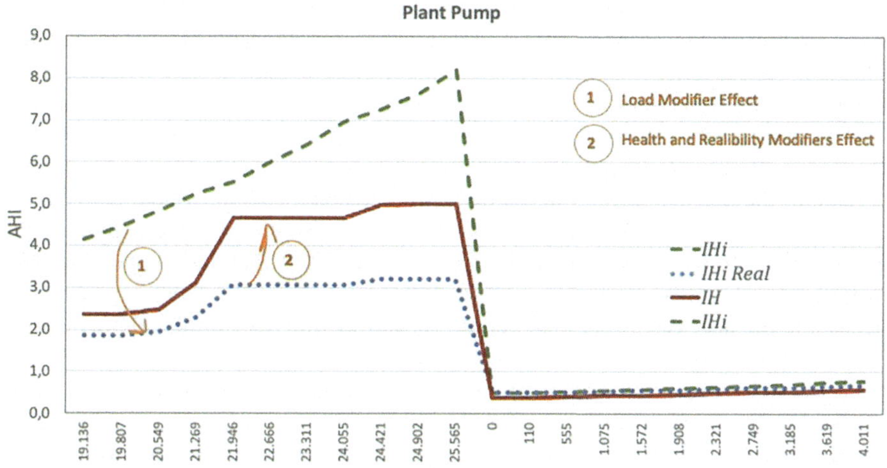

Fig. 2 Effects of modifiers on the initial health index of assets

$$\sum_{j=n}^{j=1} \gamma_j + \sum_{k=m}^{k=1} \gamma_k = \gamma \tag{9}$$

The value of γ is obtained by forcing the HIi_t to be equal to 10 (maximum limit of the HI of the asset in the model) upon reaching the estimated normal life (t_{EL}) of the asset. Therefore, $\gamma = Ln\ (Ln\ (10)/Ln\ (5.5)) = 0.301$.

The effect that the modifiers of the asset's health have in the calculation of the health index can be seen as an example in Fig. 2. Where the HIi_t is compared to the $HIiReal$ (t) and the (t). This Figure considers the possibility that the asset's health index improves with respect to the forecast (by reducing the load compared to the forecast). The figure includes the effect of the Overhaul of the pump on the mentioned indices.

3.2 *Interpretation of AHI Results*

The interpretation of Asset Health Index (AHI) can be performed in two ways: individually for decision-making on a specific asset, and comparatively for decision-making affecting a group of similar assets managed by the same organization or manager within budgetary constraints. Comparative management of AHI values is particularly valuable in global management decision-making for a network of infrastructure assets. This approach involves considering not only the AHI itself but also conducting a multivariable analysis that incorporates additional variables such as lifetime and expected degradation. By jointly managing these variables, it becomes possible to generate a comprehensive control system at the network level and effectively manage the knowledge generated. The comparative analysis between network assets is essential for enhancing the manager's ability to handle information, understand system behavior, and collect relevant data for decision-making processes. This approach enables informed decision-making and improves the overall management of the asset network.

The case study involved the assessment of eight pumps belonging to two different motor pump units, namely units A and B. This assessment allowed for a comparison of the health status of all the pumps. Based on the results obtained from the calculation of the asset health index (AHI), decisions can be made regarding the appropriate actions to take depending on the condition of each individual asset. For instance, by directly interpreting the health index values, it becomes possible to determine the necessary course of action. As an illustrative example, Table 3 showcases the potential decisions to be made based on the value of the health index. This table helps guide decision-making by linking the health index range to specific actions, such as normal maintenance, diagnostic testing, replacement planning, or immediate risk assessment and potential replacement or rebuilding. By utilizing the asset health index and the corresponding decision framework, organizations can effectively manage their assets, prioritize maintenance efforts, and optimize resource allocation to ensure the reliability and longevity of their equipment.

On the otrher hand, the model enables the establishment of decision-making guidelines through the comparison of different asset indicators and the analysis of results obtained from assets of the same class. Figure 3 presents the results, illustrating the health index of each pump on the y-axis against the operating hours since the last overhaul on the x-axis. The size of the circles represents the level of deterioration, which is measured as the deviation between the initial health index (HIi) and the final health index (HI) of each pump.

From the comparative analysis of the pumps shown in Fig. 3, several observations can be made. Pumps A2 and A4 have the lowest number of operating hours and the lowest health index. However, pump A4 has a higher health index compared to A2, despite accumulating 220 fewer operating hours. This difference is attributed to the influence of load factors and health modifiers.

Pumps B2 and B4 exhibit similar health indices, but pump B4 has accumulated approximately 3400 more operating hours than B2. Surprisingly, pump B2 shows a higher deterioration (HI-HIi) compared to B4.

Table 3 Recommendations due to AHI interpretation (De La Fuente et al., 2018)

AHI	Condition	Requirements
1 - 4	Very good	Normal maintenance
4 – 5.5	Good	Normal maintenance
5.5 - 7	Fair	Increase diagnostic testing, possible replacement depending on criticality
7 - 8	Poor	Start planning process to replace
8 - 10	Very poor	Immediately assess risk; replace or rebuild based on assessment

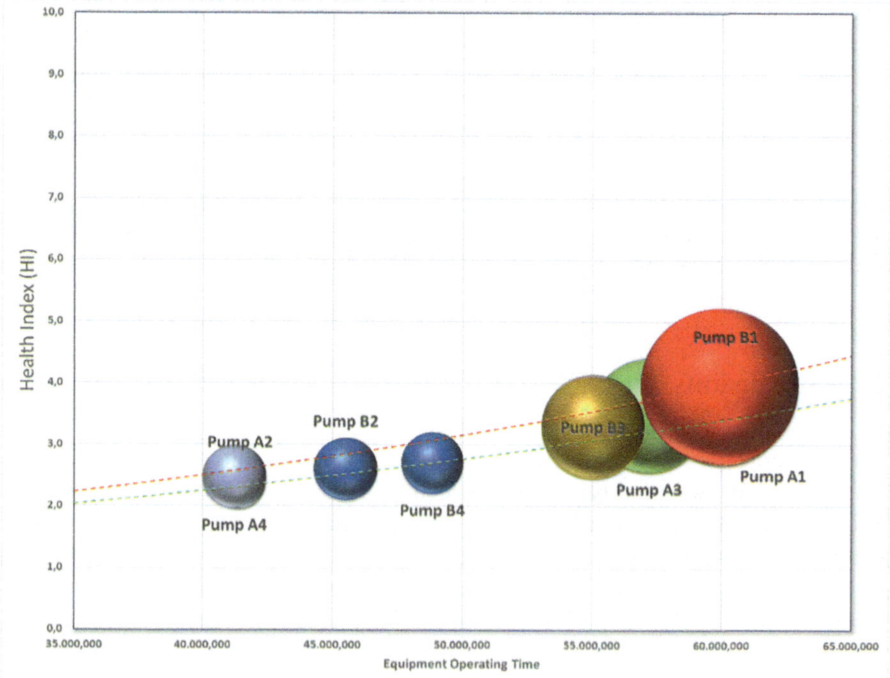

Fig. 3 Asset Health Index vs Operation Time of the pumps analysed

Although pump B1 does not have the highest number of operating hours (that distinction goes to pump A1), it has the worst health index and the highest accumulated deterioration among all the pumps analyzed.

Notably, pump B4 displays the best aging rate among the sampled pumps, while pump B1 has the worst aging rate.

As an example, pump B1 stands out as the most degraded pump among the analyzed pumps. It has accumulated approximately 59,902 operating hours since overhaul and a health index of 3.962. Despite pump A1 having slightly more operating hours than pump B1, pump B1 shows greater deterioration with approximately 200 additional operating hours.

Pump B1 is approaching the HI2 range between H = 4 and H = 6, indicating the beginning of wear signs in the asset. In case prioritization of maintenance activities is required, pump B1 takes precedence over the rest, followed by pumps A1, A3, and B3 consecutively.

By considering these observations and analyzing the comparative health and deterioration of the pumps, informed decisions can be made regarding maintenance priorities and resource allocation to ensure the optimal performance and reliability of the assets.

As an example, pump B1 is the most degraded pump from the rest of pumps analysed. This pump has accumulated approximately 59,902 operating hours since overhaul and a health index of 3.962. This pump is second only to pump A1 among all pumps in unit A and B in accumulating the most operating hours since overhaul, but shows greater deterioration than pump A1, with approximately 200 more operating hours than pump B1.

This pump B1 is close to reaching the HI2 range between the values H = 4 and H = 6, which corresponds to the period of time when the first signs of wear begin to appear in the asset. In case any maintenance activity has to be prioritised among these pumps, pump B1 will take precedence over the rest, followed by pumps A1, A3 and B3 consecutively.

4 Conclusions

In conclusion, the study conducted to calculate the asset health index of eight different assets in an industrial plant has provided valuable insights into the current condition of these assets. Although the methodology used may have certain weaknesses in terms of its mathematical formulation and consistency, it has successfully achieved its primary objective of measuring and comparing the health of the analyzed assets in an objective manner.

The collection and processing of a significant amount of location, operation, and maintenance data have enabled the estimation of the assets' current health states. This information serves as a powerful tool for the organization, providing an indicator that facilitates the prioritization of interventions, attention, and potential equipment renewal. By having a clear understanding of the assets' health status, decision-makers can allocate resources effectively, ensuring that critical equipment receives the necessary maintenance and renewal.

It is important to acknowledge the limitations of the methodology, such as the lack of consideration for the previous health index value in the calculation at each instant. This can result in flat areas in the graph, where the influence of health and reliability modifiers may not be fully captured. Despite these weaknesses, the study has delivered valuable results and provided a foundation for further improvements and refinements in future iterations.

Moving forward, it is recommended to continue exploring enhancements to the methodology, addressing the weaknesses identified, and striving for increased mathematical formulation and consistency. By refining the calculation process and considering a broader range of factors, such as the history of the asset's health index, the methodology can become more robust and offer an even more accurate representation of the asset's condition.

The complexity of the case study presents various challenges when implementing the proposed methodology. One major issue is the unavailability of historical data that spans from the commissioning of the asset to the present analysis date. This lack of historical data prevents training the model based on the actual knowledge of the asset's health status over time. However, it is still possible to execute the methodology based on the available data, as long as the current health status is known and can provide an initial approximate value.

Another challenge lies in the quality of the data itself, which can impact the accuracy of the proposed model. Inaccurate or incomplete data can lead to suboptimal results and hinder the effectiveness of the methodology. It is crucial to ensure the data's reliability and take appropriate measures to address any data quality issues that may arise.

Transferring the knowledge and experience of specialists to define indicators and modifiers for quantifying the health of the analyzed assets can be a complex task. This process requires effective communication and collaboration between experts and model implementers to ensure a comprehensive understanding of the assets and their health factors. Clear guidelines and documentation of the knowledge transfer process can help mitigate this challenge.

Despite these difficulties, implementing the proposed methodology in an organized manner allows for the generation of valid and reliable results. Validation is an essential step where the calculated health index is compared with the estimated actual health status of the assets. Expert personnel play a crucial role in this validation process, leveraging their expertise to assess the accuracy of the methodology's results. Any mismatches or discrepancies between the calculated health index and the actual health state of the asset indicate the need for recalibration.

There are three main situations that may trigger the need for recalibration. Firstly, a mismatch or discordance between the actual health of the asset and the calculated health index suggests that the methodology did not accurately assess the asset's health. Secondly, based on the acquired experience and knowledge about the asset, adjustments to the modifiers or estimated ranges may be necessary to improve the accuracy of the health index calculation. Lastly, changes in the operating conditions of the assets may require reconsideration of the modifiers' potential impact on health.

Recalibration of the model ensures that it remains aligned with the real health of the assets and takes into account any new insights or changes in the operating conditions. This iterative process of validation and recalibration enhances the accuracy and reliability of the methodology, enabling more informed decision-making in asset management.

Overall, while the case study presents challenges related to historical data availability, data quality, and knowledge transfer, a well-executed implementation of the methodology, coupled with validation and recalibration, can address these issues and yield meaningful results for effective asset management.

In summary, the study has laid the groundwork for effective asset management within the industrial plant by providing an objective measure of asset health and enabling comparative analysis. The findings serve as a valuable resource for decision-making, aiding in the prioritization of interventions and resource allocation to ensure optimal performance, reliability, and longevity of the assets. Continuous improvement and refinement of the methodology will further enhance its effectiveness and contribute to the organization's long-term success.

Acknowledgement This paper has been written within the framework of the projects INMA "Asset Digitalization for INtelligent MAintenace" (Grant PY20 RE014 AICIA, founded by Junta de Andalucía PAIDI 2020, Andalucía FEDER 2014-2020) and Geminhi (Digital model for Intelligent Maintenance based on Hybrid prognostics models), (Grant US-1381456, founded by Junta de Andalucía, Andalucía FEDER 2014-2020).

References

Australian Local Government Association. (2015). *National State of the assets report*. https://alga.asn.au/national-state-of-the-assets-report-2015/

Azmi, A., Jasni, J., Azis, N., & Kadir, M. Z. A. (2017). Evolution of transformer health index in the form of mathematical equation. *Renewable and Sustainable Energy Reviews., 76*(March), 687–700. https://doi.org/10.1016/j.rser.2017.03.094

Crespo, A. (2022). Asset health indexing and life cycle costing. In A. Crespo (Ed.), *Digital maintenance management. Guiding digital transformation in maintenance*. Springer. ISBN 978-3-030-97660-6.

Crespo, M. A., et al. (2020). Defining asset health indicators (AHI) to support complex assets maintenance and replacement strategies. A generic procedure to assess assets deterioration. In M. A. Crespo, M. Macchi, & A. Parlikad (Eds.), *Value based and intelligent asset management*. Springer. https://doi.org/10.1007/978-3-030-20704-5_4

De La Fuente, A., Guillén, A., Crespo, A., Sola, A., Gómez, J., Moreu, P., & Gonzalez-Prida, V. (2018). Chapter 126: Strategic view of an assets health index for making long-term decisions in different industries. In *Safety and reliability – Safe societies in a changing world: Proceedings of ESREL 2018, June 17–21*, 2018, Trondheim, Norway (1st ed). CRC Press. https://doi.org/10.1201/9781351174664

De la Fuente, A., Crespo, A., Sola, A., Guillén, A., Gómez, J., & Amadi-Echendu, J. E. (2021). Planning major overhaul and equipment renovation based on asset criticality and health index. In M. A. Crespo, D. Komljenovic, & J. Amadi-Echendu (Eds.), *14th WCEAM proceedings. WCEAM 2019* (Lecture notes in mechanical engineering). Springer. https://doi.org/10.1007/978-3-030-64228-0_8

Federation of Canadian Municipalities (FCM) & other seven partner organizations. (2020). *The Canadian Infrastructure Report Card (CIRC)* 2019. http://canadianinfrastructure.ca/downloads/canadian-infrastructure-report-card-2019.pdf

GB DNO groups. (2017). DNO common network asset indices methodology. Health & Criticality – Version 1.1. Available at: https://www.ofgem.gov.uk/system/files/docs/2017/05/dno_common_network_asset_indices_methodology_v1.1.pdf

Hjartarson, T., & Otal, S. (2006). Predicting future asset condition based on current health index and maintenance level. In *ESMO 2006 – 2006 IEEE 11th international conference on Transmission & Distribution Construction, operation and live-line maintenance*. IEEE.

López, K., de Calle, S., Ferreiro, A. A., & Sierra, B. (2019). Dynamic condition monitoring method based on dimensionality reduction techniques for data-limited industrial environments, 103114. ISSN 0166-3615. *Computers in Industry, 112*. https://doi.org/10.1016/j.compind.2019.07.004

Naderian, A., Cress, S., Piercy, R., Wang, F., & Service, J. (2008). An approach to determine the health index of power transformers. In *Conference record of the IEEE international symposium on electrical insulation* (pp. 192–196). IEEE. https://doi.org/10.1109/ELINSL.2008.4570308

Naderian, A., Piercy, R., Cress, S., Service, J., & Wang, F. (2009). An approach to power transformer asset management using health index. *IEEE Electrical Insulation Magazine, 25*(2), 2. https://doi.org/10.1109/MEI.2009.4802595

Scatiggio, F., & Pompili, M. (2013). Health index: The TERNA's practical approach for transformers fleet management. In *IEEE electrical insulation conference (EIC)* (pp. 178–182). IEEE.
Serra, J., De la Fuente, A., Crespo, A., Sola, A., Guillén, A., Candón, E., & Martínez-Galán, P. (2019). A model for lifecycle cost calculation based on asset health index. In *International Conference on Smart Infrastructure and Construction 2019 (ICSIC)* (pp. 91–98). ISBN 978-0-7277-6466-9. Available at: https://www.icevirtuallibrary.com/doi/10.1680/icsic.64669.091
Silvestri, L., Forcina, A., Introna, V., Santolamazza, A., & Cesarotti, V. (2020). Maintenance transformation through industry 4.0 technologies: A systematic literature review. *Computers in Industry, 123*. ISSN 0166-3615. https://doi.org/10.1016/j.compind.2020.103335
Vermeer, M., Wetzer, J., van der Wielen, P., de Haan, E., & de Meulemeester, E. (2015). Asset management decision support modeling, using a health and risk model. *2015 IEEE Eindhoven PowerTech*, 1–6. https://doi.org/10.1109/PTC.2015.7232556

Eduardo Candón Fernández is an Industrial Engineer from the School of Engineering of the University of Seville, member of INGEMAN (National Association for the Development of Maintenance Engineering in Spain) and member of the SIM (Intelligent Maintenance Systems) research group of the Department of Industrial Organization and Business Management of the School of Engineering of Seville, and currently a PhD student in the program of mechanical engineering and industrial management.

Adolfo Crespo Márquez is currently Full Professor at the Department of Industrial Management, School of Engineering of the University of Seville. He holds a PhD with Honors in Industrial Engineering from this same university. He is included in the Stanford University's list of world top-scientists in 2023 and he received the Spanish Maintenance Award from the Spanish Maintenance Association (AEM) in 2020. With research works published in many top journals, he is the author of 8 books and editor of 6 books with Springer, Aenor, IGI Global and INGEMAN, about maintenance, warranty, supply chain and assets management. Prof. Crespo is Fellow Director of ISEAM, Editor-in-Chief of EAMR Springer Book Series and led the Spanish Research Network on Assets Management and the Spanish Committee for Maintenance Standardization (1995–2003). He also leads the SIM research group related to smart maintenance systems and has extensively participated in many engineering and consulting projects for different companies, for the Spanish Departments of Defense, Science and Education as well as for the European Commission (IPTS). He is the President of INGEMAN (a National Association for the Development of Maintenance Engineering in Spain) since 2002.

Antonio Jesús Guillén López obtained his PhD in Industrial Engineering from the University of Seville, currently holds the position of Professor in the Department of Organization of Companies at the University Complutense of Madrid. He is member of the SIM (Sistemas Inteligentes de Mantenimiento) research group at the University of Seville, contributing significantly to the field of intelligent maintenance systems. His career has been dedicated to impactful research, particularly in the areas of asset management and maintenance. His leadership extends to serving as the Secretary General of ESReDA since 2022, overseeing European Safety, Reliability, and Data Association initiatives. Since 2012, he has collaborated as a consultant with INGEMAN (Asociación para el Desarrollo de la Ingeniería de Mantenimiento) on projects with notable companies, including ADIF, ENAGAS, TALGO, AZVI, VIESGO, and ABENGOA.

General Bases to Hierarchy Definition for Digital Assets in Railway Context

Mauricio Rodríguez, Adolfo Crespo Márquez, Antonio Jesús Guillén López, and Eduardo Candón Fernández

Abstract Defining the existence of a digital asset, integrating multiple platforms that represent its entities digitally, and simultaneously meeting the specific demands of the operational context of railway infrastructure systems represents an unresolved challenge for this industry. This study focuses on the search for commonalities, complementing the perspectives of the scientific community and research centers with real-world applications. From there, the development of a framework presented in our research emerges, capturing both the state of the art and practice, providing a starting point for the development of scientific discussions and the search for future models that offer an effective solution to the problem. The integration of maintenance management models with architectures for the development of digital twins in Industry 4.0, and the applied study of the railway industry itself, are part of the foundation of this study. Seeking to adhere to the principles already proposed for Industry 4.0, the scheme introduces new relationship factors that will be prototyped in the industry, especially in railway infrastructures, allowing for scalability and the digitization of processes as crucial as the criticality assessment for asset prioritization.

Keywords Railway maintenance · Asset management · Criticality analysis · Asset hierarchy definition · Industry 4.0 · Digital twin

1 Introduction

The principal motivation to develop this research is to provide the scientific community and the industrial world with a comprehensive framework to initiate the standardization of digitalization in the railways industry. In pursuit of this goal, the

M. Rodríguez (✉) · A. Crespo Márquez · A. J. Guillén López · E. Candón Fernández
Universidad de Sevilla, Sevilla, Spain
e-mail: maurodher@alum.us.es; adolfo@us.es; ajguillen@us.es; ecandon@us.es

A. Crespo Márquez et al. (eds.), *Advances in Asset Management: Strategies, Technologies, and Industry Applications*, Engineering Asset Management Review 3, https://doi.org/10.1007/978-3-031-52391-5_6

authors contribute their invaluable experience in real applications, skillfully integrated with state-of-the-art practices for the 4.0 industry.

This paper aims to establish a robust hierarchical structuring scheme for railway assets, taking into consideration a systemic, holistic, and digital perspective. The framework encompasses three fundamental dimensions that are pivotal to the success of the digital transformation: the real-world dimension, which pertains to the physical assets in the railways (Zheng et al., 2021); the digital dimension, manifested in multiple platforms seeking to generate their own digital elements (Schweichhart, 2016); and the management dimension, derived from well-established models like MGM (Crespo Márquez, 2007), ISO (55,000), and UIC standards, all geared towards maximizing value and streamlining processes to achieve desired outcomes.

One of the primary challenges confronting the industry lies in selecting an appropriate Maintenance Management model to efficiently oversee assets. It is crucial to consider both the outputs and inputs of the chosen model from the outset. The definition of inputs and the way they are integrated will significantly influence the success of implementing any management model, particularly at higher levels of decision-making ("the last layer") (Schweichhart, 2016). Thus, understanding the entire journey from start to finish is essential before embracing any technological or digital solution, no matter how promising they might appear in addressing maintenance issues.

Meanwhile, the omnipresence of digitalization in various OT (Operational Technology) and IT (Information Technology) platforms poses its own set of challenges. Aligning these diverse solutions to converge and add value to the process, rather than causing confusion or entropy, becomes a critical endeavor. Currently, there lacks a widespread consensus on a general architecture that can cater to the demands of Industry 4.0 for networked systems, specifically within the context of railway infrastructure systems (UIC A.W, 2022).

This research sets out to bridge the gaps in these dimensions and overcome the challenges through a comprehensive and systemic approach. By proposing a hierarchical framework that incorporates the real-world, digital, and management perspectives, it paves the way for a standardized approach to digitalization in the railways industry.

It is evident that standardizing digitalization in the railways sector will unlock numerous benefits. Streamlined processes and improved asset management will lead to enhanced efficiency, reduced downtime, and overall cost savings. Additionally, the adoption of Industry 4.0 principles will empower railways to stay competitive in an increasingly digital world.

2 Research Methodology

The methodology employed for this study encompassed three key approaches: a comprehensive review of scientific and technical literature in European research, consultations with infrastructure managers across Europe, and the examination of practical experiences related to the development and implementation of a management and digitalization model within one of the European railway systems.

In the initial phase, the focus was on conducting a bibliographic and bibliometric review of scientific publications, utilizing the Scopus database as a reference. The primary objective was to identify the level and intensity of the relationship between the three main concepts under study: Hierarchy of Assets, Digital Twin, and Railway. The search pattern employed was "railway" AND "digital twin" OR "railway" AND "hierarchy."

Subsequently, the second phase concentrated on investigating in railway research centers and ongoing projects to identify potential models in development or experimentation that could serve as a foundation or contribute inputs to the proposed model. This phase involved examining the technological and railway regulation systems pertaining to digitalization, hierarchy, and maintenance. Additionally, research was conducted on digitalization projects within European railways, including their respective internal forums, to gain insights into the industry's real-life practices.

3 Synthesis Review

3.1 Technical Scientific Literature

This review aims to provide a fundamental vision of the art state in which the discussion of these concepts within the scientific community is located, and the evolution over time of them. The results obtained, presented as number of publications over time are shown in 1a and the bibliometric study developed through the VOS viewer platform, allows to appreciate Figs. 1a and 1b, both the intensity of the concepts, as well as their relationship and their connection with other concepts treated in the analyzed studies. Out of a total of 668 results obtained from SCOPUS.

The analysis indicates that within the railway context, the concepts of digitalization, digital twins, and Industry 4.0 are still in the nascent stages of development. Despite the utilization of the Analytic Hierarchy Process (AHP), which is predominantly associated with risk assessment processes, its application to hierarchize railway assets and the develop of their corresponding digital twins remains limited. This being precisely the "Gap" identified as our object of study.

Another observation from the bibliometric review is that over the past 5 years, the subject has only begun to emerge in connection with Geographic Information Systems (GIS). However, this emergence is confined to specific contexts with his

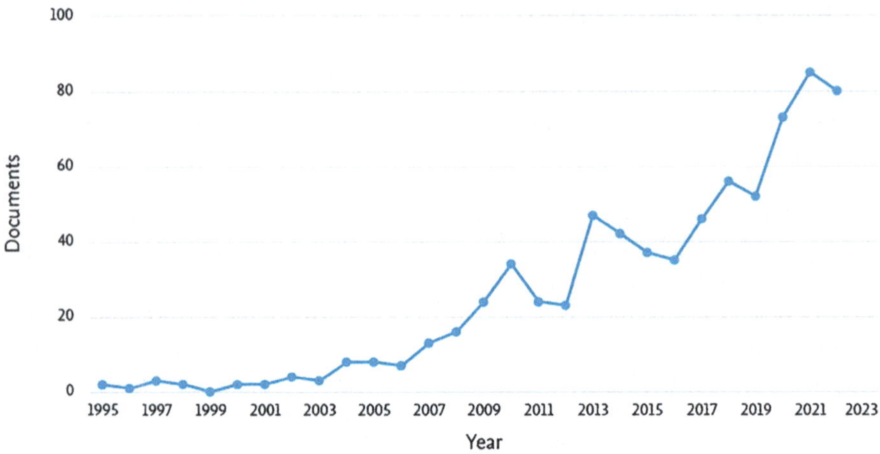

Fig. 1a Evolution 1995–2022 on SCOPUS of mains concepts

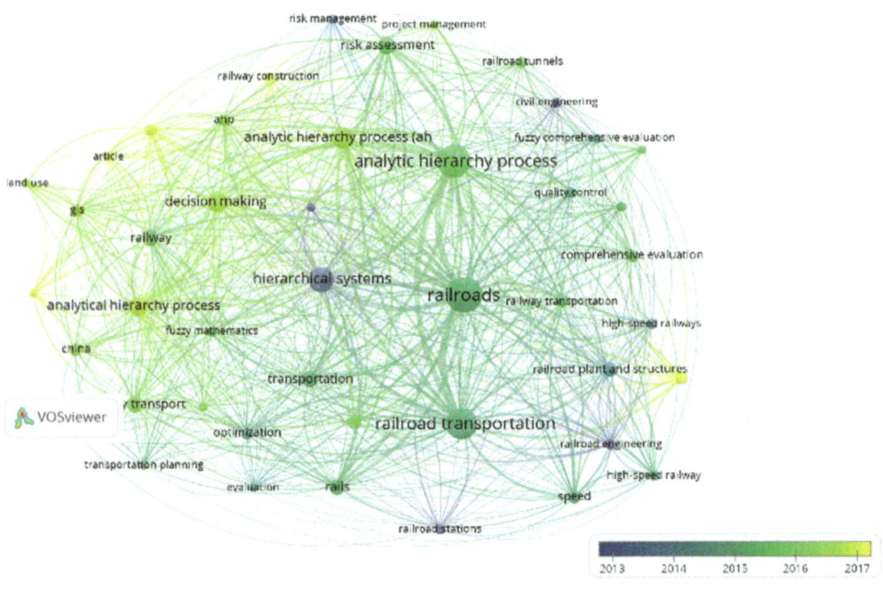

Fig. 1b Evolution, Intensity and Relationship of mains concepts

own rules and does not offer a comprehensive view of the hierarchy of digital assets within the railway industry.

Then, from the existing literature review is possible to see ample opportunities for the development of models and the initiation of scientific discussions on the subject matter. It is worth mentioning, neither the hierarchy nor the taxonomy of railway infrastructure assets appear to be regulated by any scientific model yet.

Rather, they have been derived from practical considerations implemented individually by the railway operators themselves.

In summary respect of the actual scientific discussion on these aspects, we confirm our gap objectives. We can see that as the industry moves towards digital transformation, the need for a coherent and scientifically grounded framework for hierarchizing railway assets becomes increasingly apparent. This reveals that is necessary calls for further research and scholarly dialogue to address the existing gaps and establish a unified approach to the integration of digitalization, digital twins, and Industry 4.0 principles in the railway domain.

3.2 Research European Projects

Currently there are multiple efforts to incorporate digitalization in the various railway entities, giving rise to several initiatives both from the scientific and practical perspective. The approach from the scientific point of view as analyzed in 3.1 suggests which respect of hierarchization of asset in the railway digitalization context, the discussion is weak, so it is of special attention to complement it with the applied visions. Aiming to rescue the praxis from the industry itself, these 3 sources have been investigated, corresponding to applied research projects, with a strong link to the current railway ecosystem:

- Europeans Investigations: Shitt2rail (Rail, s.f.) Fig. 2: Shift2Rail R&I Programme & Projects. "The vision of the Shift2Rail is to deliver, through railway research and innovation, the capabilities to bring about the most sustainable, cost-efficient, high-performing, time driven, digital and competitive customer-centred transport mode for Europe".
- Industry-Specific Research: UIC (UIC, s.f.): "The UIC research portal is intended to play a crucial role in facilitating this process. This portal collects and maintains information from numerous global sources. Its primary purpose is to build on information shared by his members and research institutes and by industry-leading research providers used and recommended by his members, as well as data obtained by linking up globally with other rail research databases".
- Railway Innovation Hub (RIH, 2023): His mission is to promote railway technology and knowledge at international level through the generation of collaborative R&D projects, the commercialization of technology and know-how, the promotion of entrepreneurship and the provision of specialized services.

From the multiple research project develops in these 3 instances, we rescue 2 studies specifically related with the Gap proposed un our investigation, the first of them referring to the taxonomy of assets developed by "The Railway Innovation Hub, Spain" and the second to the study of "Big data for Asset Management", developed by UIC. From where the following key aspects for this research are rescued:

IP1	IP2	IP3	IP4	IP5	CCA
TD 1.1 - Traction Systems demonstrator	TD 2.1 - Adaptable communications for all railways	TD 3.1 - Enhanced Switch & Crossing System Demonstrator	TD 4.1 - Interoperability Framework	TD 5.1 - Fleet Digitalization and Automation	WA 1 - Long-term needs and socio-economic research and system platform demonstrators
TD 1.2 - Train Control and Monitoring System Demonstrator	TD 2.2 - Railway network capacity increase	TD 3.2 - Next Generation Switch & Crossing System Demonstrator	TD 4.2 - Travel Shopping	TD 5.2 - Digital Transport Management	WA 2 - KPI method development and integrated assessment
TD 1.3 - Carbody Shell Demonstrator	TD 2.3 - Moving Blocks	TD 3.3 - Optimised Track System	TD 4.3 - Booking & Ticketing	TD 5.3 - Smart Freight Wagon Concepts	WA 3 - Safety, Standardisation, Smart Maintenance, Smart Materials & Virtual certification
TD 1.4 - Running Gear Demonstrator	TD 2.4 - Fail-Safe Train Positioning	TD 3.4 - Next Generation Track System	TD 4.4 - Trip Tracker	TD 5.4 - New Freight Propulsion Concepts	WA 4 - Smart Mobility (4.1 Smart Planning, 4.2 I2M)
TD 1.5 - Brake Systems Demonstrator	TD 2.5 - On-board Train Integrity	TD 3.5 - Proactive Bridge and Tunnel Assessment, Repair and Upgrade Demonstrator	TD 4.5 - Travel Companion	TD 5.5 - Business analytics & implementation strategies	WA 5 - Energy and Sustainability (5.1 Energy, 5.2 Noise and Vibration)
TD 1.6 - Doors and Access Systems Demonstrator	TD 2.6 - Zero on-site testing	TD 3.6 - Dynamic Railway Information Management System (DRIMS) Demonstrator	TD 4.6 - Business Analytics Platform		WA 6 - Human Capital
TD 1.7 - Train Modularity in Use	TD 2.7 - Formal methods and standardisation for smart signalling systems	TD 3.7 - Railway Integrated Measuring and Monitoring System (RIMMS) Demonstrator	TD 4.7 - Overall IP4 Coordination and Demonstrations		
TD 1.8 - Heating, Ventilation, Air conditioning and Cooling (HVAC)	TD 2.8 - Virtually-Coupled Train Sets	TD 3.8 - Intelligent Asset Management Strategies Demonstrator (IAMS)			
	TD 2.9 - Traffic management system	TD 3.9 - Smart Power Supply Demonstrator			
	TD 2.10 - Smart radio-connected all-in-all wayside objects	TD 3.10 - Smart Metering for Railway Distributed Energy Resource Management System Demonstrator			
	TD 2.11 - Cyber Security	TD 3.11 - Future Stations Demonstrator			

Fig. 2 Shift2Rail R&I Programme & Projects 2022–23

1. From RIH, in the lasts years efforts on the development of BIM Railway Classification System Manual. Focused mainly on BIM systems (Ali et al., 2022) and consequently on digital models for construction only.
2. From S2R, is reaffirmed that it is possible to find multiple investigations that point to the digitalization of assets, but that most of them focus on rolling stock and its peripherals, leaving as a corollary that in the aspect of digitalization of railway infrastructure is limited to a few.
3. From UIC research, we found a very interest approach, in (Roda, I, 2022) it is argued that it is essential to adopt a systemic perspective for the management of

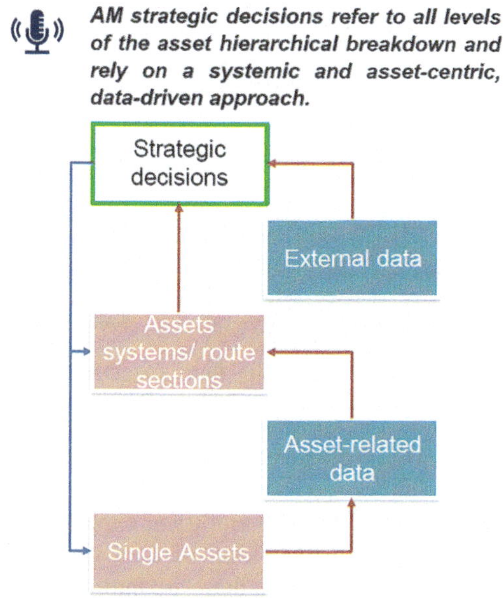

Fig. 3 Top-down and bottom-up approaches for systemic and asset-centric, data-driven decision-making. (Roda, 2022)

physical assets (Macchi et al., 2012), which will certainly condition the way in which data is handled (big data management), because it is these same, coming from the individual asset (bottom up), which determine the strategies (strategic decision-making) that must be addressed systemically (top down). Fig. 3.

4. In (Roda, 2022) the existence of a high-level taxonomy (Sedghi et al., 2021), characteristic of railway systems, is recognized, where it is possible to recognize 2 large groups of assets: network assets, discrete assets see Fig. 4.
5. In the consensus of the studies in UIC it is stated that Is necessary a methodology for the adoption of Big Data, where it is indicated that one of the first activities will be the definition of "relevant elements", his impact on the business and in the decisions may be exist over itself. Where for each of them "the relevant element" there will be the "data wish list" necessary to make those decisions better.

We can indicate from this study that these aspects, as a whole, provide a first approach to railway 4.0 industry, and together justify the need to have a standardized architecture, which allows the exploitation of data in a systematized way.

3.3 *Railway Administrators*

The third part of the research has been developed on the actual application of a maintenance management model for a railway operator in Europe. That it has considered the revision of its current data models and applications that allow its exploitation. In this context, because of an internal inquiry at European level, it has been

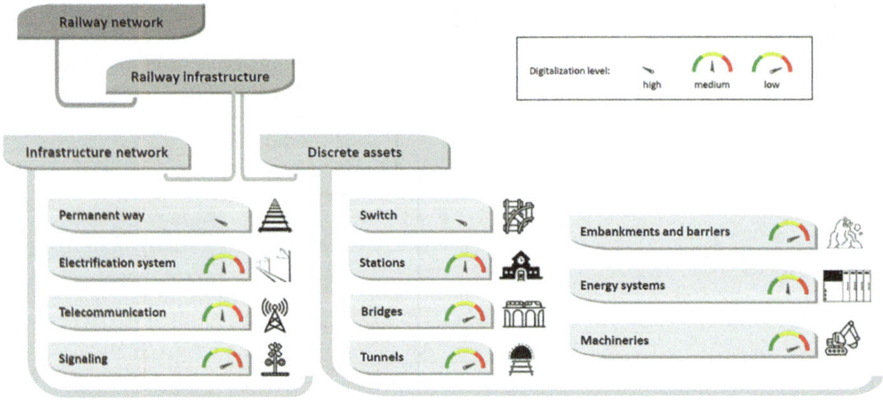

Fig. 4 Asset taxonomy for railways organization and current digitalization levels (Roda, 2022)

possible to collect the empirical experience of several operators, which as a result demonstrates a high level of data variability in other railways. This experience allows us to verify that there are multiple models that have developed naturally over time, that is, they have been adaptations to a progressive growth of information, usually hyper specialized, which effectively solve their local problems, but wich do not have systemic capacities that allow the exploitation of data and the identification of assets in a digital and unique way. In the case study itself, we have found that such a level of personalization becomes an advantage from the perspective of a specialty and its own requirements, but that, on the other hand, it becomes a great disadvantage when it comes to exploiting the information, given the architecture of the system and the data. Problems such as taxonomy, multiple identification of the same asset, multiplicity of structures and / or hierarchies, multiplicity of information sources (ranging from databases to pdfs un-processable digitally), in short, the "big data" ends up transforming into "Frankenstein data". From the applied experience, at least 3 factors that directly affect digitalization are recognized:

1. Reference system: an aspect in constant discussion from the railway point of view. Given their nature, railways began to reference their assets according to tracks (lines) and points on them, however, the complexity of growth has led the system rather to a network model, composed of nodes and strokes that together with a certain topology become a complex system, recognizing that the current system is insufficient to meet the needs of digitalization. In this sense, an issue to be resolved when considering a hierarchy of assets is the reference system itself.
2. Attributes Heterogeneity: each type of asset is characterized by having its own set of attributes according to the specialty to which it provides service, this condition has led computer developers in the railway world to propose ultra-specialized data solutions, which prevent the standard characterization of an element or digital entity, consequently, they do not facilitate their relationship on a common basis that allows their comparison and respective hierarchization.

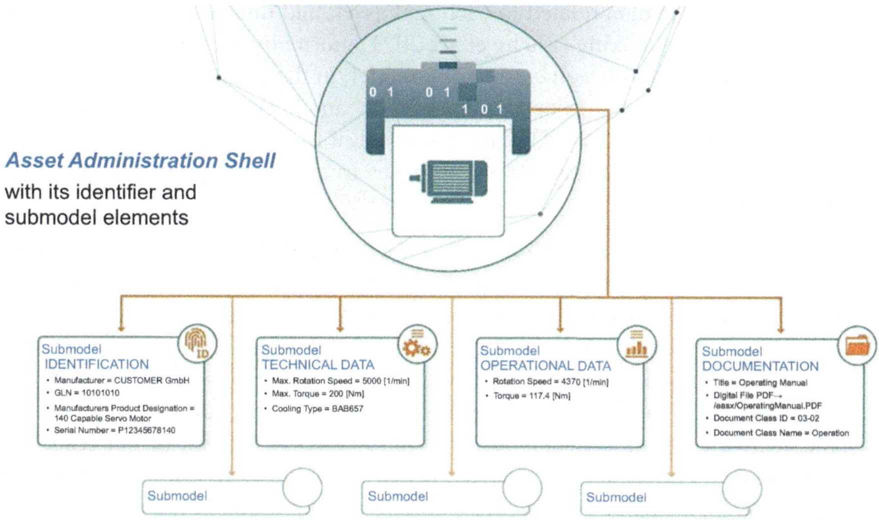

Fig. 5 Asset Administration Shell and its submodel (Bader et al., 2018)

3. Information Sources: Most railway operators in Europe have ownership over infrastructures, which implies that at least 3 different functional contexts coexist (Railway Operation, Construction of new networks, Conservation and Maintenance of Infrastructures), on which assets also coexist, but respond to very different objectives, which makes each context separately define its own data structure and information systems for the exploitation of them, granting multiple different identities to the same physical asset, which makes the data unmanageable from the computer perspective, forcing to create fictitious relationships that allow interpreting a set of data, but that do not reflect exactly reality. This drives the need for data integration, throughout the asset lifecycle, where the "digital physical asset" is unique, in a single functional structure and on which the layer or Asset Administration Shell, see Fig. 5, is placed (Bader et al., 2018) that characterize it in one context or another.

4 Conceptual Scheme for Development a Digital Architecture: Railway Application

The work presented in this paper develops an adaptation of the current models in the scientific discussion about digital twins and 4.0 industry. Adding elements and their relationship in a multi-system context (Macchi et al., 2012), applied specifically to the context of the railway industry. This provides as a result a proposal that is based on the 3 groups of layers indicated at the introduction, the first that is related to the real physical assets, the second that groups all the capture, processing, modeling and integration of the data related to each real asset, and finally the third layer that consolidates decision making (Zheng et al., 2021), normally supported in multivariable decision models, such as the criticality model, in the case of asset hierarchy.

As an approximation to a general data model (Candón et al., 2019) for the development of the railway industry 4.0, and as a result of the practical application in the case studies, a model is presented that proposes to correlate at least 5 elementary dimensions, these are:

1. Hierarchical Structure of the assets which will establish their functional dependence for the required purpose of the set Functional Units of System. This hierarchical decomposition may be developed according to the regulations that apply (if any) to each specialty. Typically, systems, subsystems, equipment, and components will be recognized (Fig. 6).
2. Units Functional of System, recognized as the minimum unit of process that generates value. This concept, which is introduced as a result of our research, results from a convenient convention for the grouping of assets in a networked distributed linear asset environment. It is very important to note that what is called "value" is closely related to the purpose of the business. Aligning with it the systemic perspective that is required for decision making, as we have cited in Sect. 3.2.

   To exemplify the above, consider the case of a train that must travel from station A to station B, its purpose is to arrive on time and fulfil that journey by transporting a certain amount of passengers or merchandise. If any of the systems serving that route fails, the purpose cannot be executed, in this case we will recognize a Network UFC. Other case may be exemplified with UFS of a discreet nature, that is, they are located at a point or node of the network and that

Fig. 6 RAMI 4.0 Functional Architecture (Schweichhart, 2016)

they fulfil a specific purpose, as is the case of level crossings, which complement the railway operation have as their objective the protection of railway and road safety, in this case if there is any failure in the safety barrier, this discreet UFS will no longer fulfil its purpose.

3. Real Assets Layer, which uniquely identifies each physical asset. In this layer you can recognize the physical assets that make up a UFS, for example in the railway case are Rails, sleepers, fixings, Switch & Crossing, Track circuits, Interlockings, catenary wire, posts, transformation centers, bridges, tunnels, etc.
4. Digital Twin layer, which digests, recognizes and processes the data of each real Asset, providing the Digital identity and some result according to some given output according to the ontology (Li et al., 2022) defined for each case. In (Malakuti et al., 2020) is define a discrete digital twin like a single entity that provides value without needing to be broken down further. For example, the gearbox or motor for a ball mill in mining can be monitored and reported on at this entity level. Assembling discrete digital twins to create a composite digital twin is shown in Fig. 7 as a vertical expansion that describes the increase in composition from a single to many entities.

   A composite digital twin is a combination of discrete digital twins that represent an entity comprising multiple individual components or parts. The composition may take place at different levels. For example, a production cell is a composite entity, whose digital twin consists of the digital twins of the devices within the production cell. An entire plant is a system, whose digital twin consists of several others composite digital twins.
5. BDM Layer, Layer of business decisions from the Big data, for example an operational risk assessment of railway infrastructure (Weik et al., 2022) of the system, which includes not only the health data of the assets but also the opera-

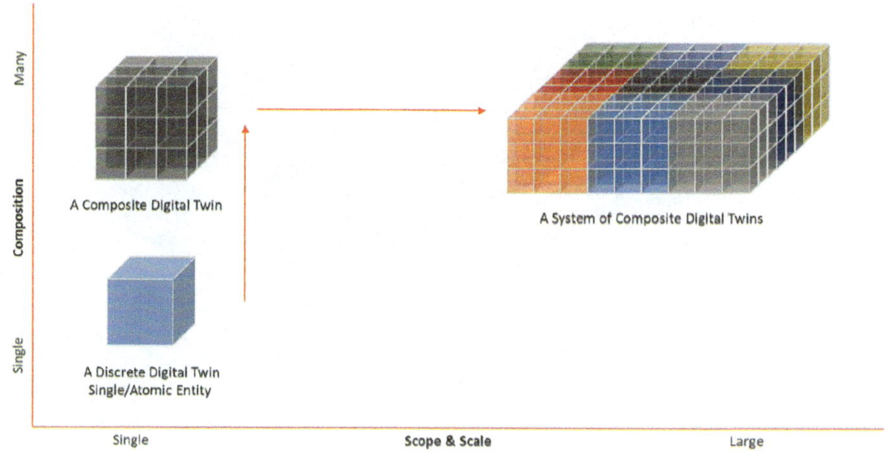

Fig. 7 Composite Digital Twin. (Malakuti et al., 2020)

tional possibilities that affect them, allowing to establish a risk management model using Markov chains, on which business decisions could be made.

In general, this layer produces the information from the business rules and exploitation of information that the Stakeholders demand.

4.1 RDTA: Railway Digital Twin Architecture (Schematic)

Based on the principles proposed in the digitalization model structures such as RAMI, CDT among others (Zheng et al., 2021), the scheme shown in Fig. 8 introduces a new concept that allows it's adaptation to the railway industry, this takes charge of the way in which the minimum units of value are defined in a railway context, which will be classified as both network units or discrete units (Leitner et al., 2017). We have called this concept a Functional System Unit.

Therefore, the integration of a real asset, duly structured in a hierarchy that places it at a certain level and a functional unit in which it provides a service, all on the layer of digitization rules (recognition of the same asset in multiple systems),

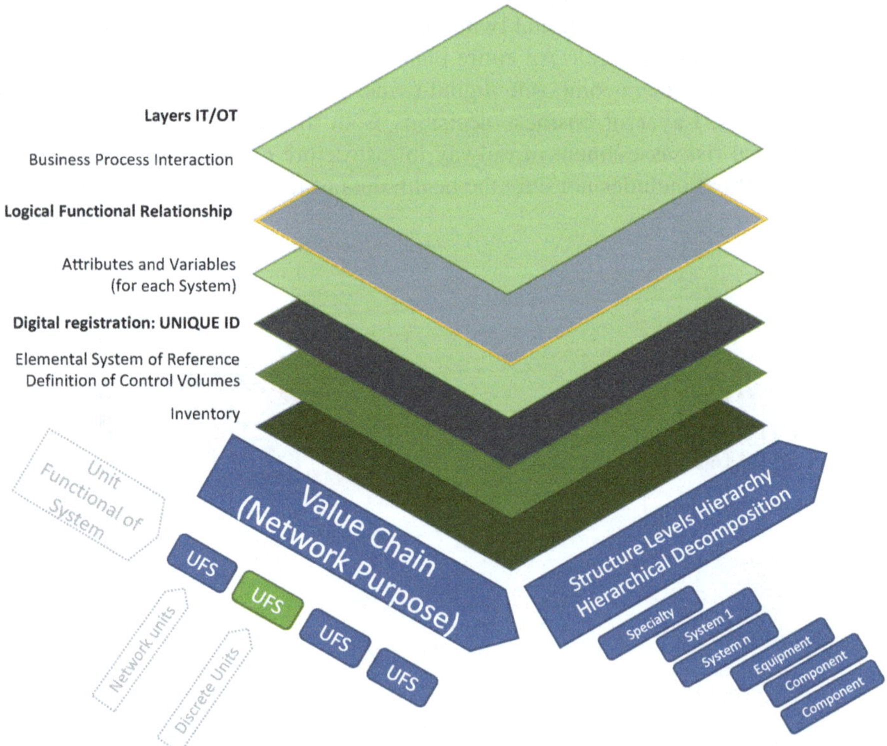

Fig. 8 Railways digital twin architecture

will provide each asset with what we have called a digital identity, which can then be conveniently exploited in the last layer of business decisions. Those that may be to a lesser or greater degree digitalize them, according to the internal digitalization models (Weik et al., 2022) that each architecture implementation defines, being able to be these simple relationships between databases, up to complex Machine Learning models integrated in real time.

UFS: Unit Functional of System, minimum unit within the value chain, where value is added to the process as such, autonomous unit where all the existing systems in the hierarchy coexist, which will be related, logically and functionally through these UFS, allowing to establish a base relationship parallel to the hierarchy of systems / equipment / component, that allows to emulate and assess the function of each asset with it's impact at the systemic level (Mohammadi & El-Diraby, 2021), the above will complement the traditional structure / hierarchy of assets and will cross the layer of inventories, allowing to order from the base to the assets according to their taxonomy and their function provided at the same time. Recognizing, also, at this level two types of units, the "Network units" and the "Discrete Units" For example, a journey vs a bridge, a tunnel. (Carretero et al., 2003).

HIERARCHY LEVELS: Defined as System, subsystem, component according to the railway taxonomy.

LAYERS: Grouped into 3 macro levels, real level, digital level, decision level.

Real Level: deals with the physical assets, refers to the physical and tangible aspect of the railway industry. It encompasses all the physical assets and infrastructure that make up the railway system, such as tracks, trains, signaling equipment, bridges, tunnels, and stations. This level deals with the actual, on-ground components that ensure the smooth functioning of the railway network.

Digital Level: creates virtual representations through digital twins. The Digital Level is the realm of virtual representation and data integration. It involves the creation of digital twins for each physical asset at the Real Level. A digital twin is a virtual replica of a real-world asset, capturing its characteristics, behavior, and performance through data. In the Digital Level, various computer systems, such as Enterprise Asset Management (EAM), Building Information Modeling (BIM), and Geographic Information Systems (GIS), play a crucial role. These systems collect, process, and integrate data related to each asset, providing a comprehensive and dynamic digital representation.

Decision Level: utilizes data-driven insights to make informed and strategic decisions for the industry's success. Asset management focuses on maintaining and optimizing the physical assets to ensure safe and efficient operations. This includes regular maintenance, inspections, and repairs to prevent failures and minimize downtime.

4.2 Proof of Concept: Railway Application

The study on computer solutions for Asset Management EAM, BIM Systems, GIS Systems, allows us to recognize that it is feasible at the digital level to recognize an asset through a volume of control (real, digital or both), which may be as small or as large as the user defines. Taking this into account and the data of a real railway system, a sequence of steps is proposed that should at least be considered for the implementation of the scheme proposed in Sect. 4.1, our objective, under the framework of the digitalization of assets for decision-making regarding their management and maintenance, in line with the framework proposed by (Crespo Márquez, 2022).

Step 1: Considering the criteria established in our RDTA scheme, we define a digital asset within a control volume, which will have as an elementary reference geo spatiality, recognizing this as the most absolute feasible reference, each control volume will therefore have it's 3 dimensions, expanding the current reference systems that are only limited to flat referencing (v-pk). To make that, as a proposal proof of concept we consider all the assets existent into inventory (rail, sleepers, signally, welding, swishes, etc.) that belong to some functional section (for example from station A to station B), then propose the relationship with the element defined an BIM platform to identify and match the singularity elements defined on BIM model with the individual element defined on the inventory platform. Finally, we relate the 2 previous platforms and correspond data references with the GIS system, is very important to remark that the systems currently are not integrated. Figure 9. So to extend the model requires the creation of individual digital entities capable of being recognized in interpreted in each of the systems from which they are demanded, this will imply the creation of essential attributes for each asset and then

Fig. 9 Example of Real word Railways Assets vs Digital word on BIM, EAM or GIS system (Image: Siemens Mobility)

in your AAS all the data models that are needed will be added, starting with the hierarchy itself that from our perspective is considered as an essential data in given entity.

Step 2: Categorization of functional units (network or discrete). To define if a unit is a network or discrete unit, we concentrate in the function "the proposal" of the UFS, in this sense we can define it like an example of network unit: the segment between 2 stations, including all kinds of assets that belong to this segment (permanent way, electrification system, signaling, telecommunications, etc.). For the other side if we find a bridge, a tunnel or switches and crossing elements, we are extracting them from the network unit and recognize like a discrete unit, with a proposal itself.

Step 3: Hierarchical structuring of assets, defined by their taxonomy according to specialty. Is a principal effort to give a digital identity to the assets, mainly on the EAM systems, where we can give the parameters for the decision-making model.

Step 4: Allocation of the real assets into each UFS with it's respective hierarchical structure associated, according to how many real assets you have, level of grouping by specialty connected online. Level of detail of each specialty according to each Linear or point element defined. In this point it is very important to define some rules to truncate or divide the asset, mainly the linear assets, because for example: not necessarily the limit or the end of one permanent way is the same limit or end of a telecommunication system, as seen in Fig. 10. Then the final of the UFS will not be referenced to some unique linear reference system, but to multiples systems, as seen in Figs. 11 and 12, integrated in a digital twin.

Like a result of the application of the model we observe, first that if the organization aligns the digital definitions of each asset through one "management perspective", according to one model, allowing all the uplevels for the exploitation of the information to decision-making layers. The homogenization and standardization of each unit with the same criteria, allow too to scale the model for all the company (Fig. 12), gives the capability to use advanced techniques of data mining or business analytics to get the best result for the business, always aligned with one criticality perspective.

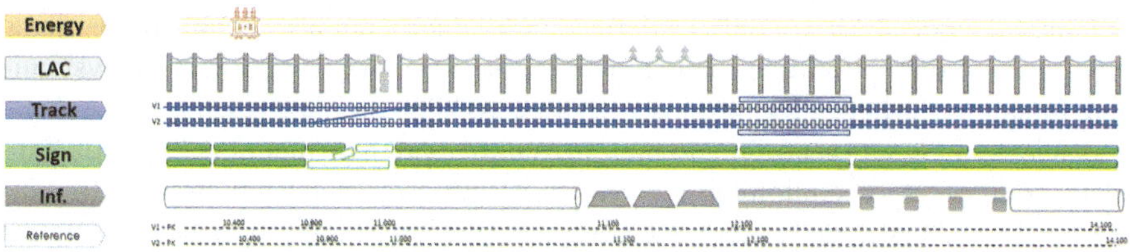

Fig. 10 Example of coexistence of complex multiple end-limit between specialities on a railway network

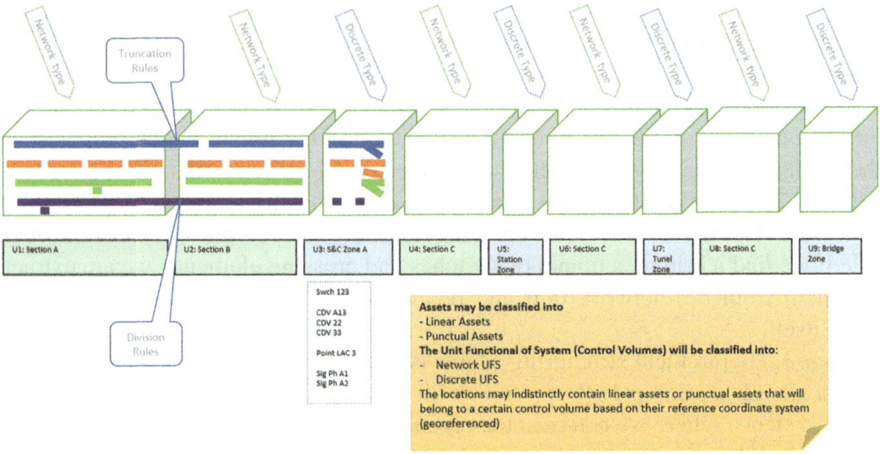

Fig. 11 UFS Railways Multiply System Scheme

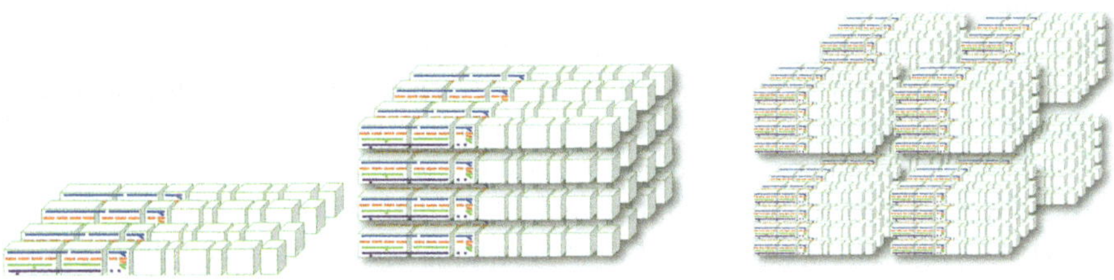

Fig. 12 Business, Network and Enterprise Units

5 Conclusions

The main contribution of this study is to open the scientific discussion around the establishment of two clear lines of research and development for the future structuring of digitalization models applied to the railway industry.

The first line involves the incorporation of the "management" factor through a model that aims to maximize the business results. While digitalization has predominantly focused on technical aspects, this proposal seeks to integrate all computer models under a clear Management model (The MMM). This approach recognizes that digital assets must not only be defined individually but also establish functional relationships that determine their impact on the business. By adopting a systemic and functional criteria, the proposed model, RDTC, emphasizes the hierarchical dependence of assets according to their taxonomy and their functional relationships (data source) from equipment to system (Button-up perspective). Finally this approach enables decision-making (top-down) with the appropriate information about all assets and their overall impact.

The second line, is the model RDTC itself like a conceptual scheme, incorporating the systemic perspective in the asset management models, consequently, of their digital twins. The digital asset normally is defined individually, but not necessarily establishing a functional relationship, that determines the impact on the business. In this sense, it is proposed as an essential factor, the use of a systemic and functional criteria, which fixes the hierarchical dependence of the assets respect to their taxonomy, and their functional relationship (data source) from the equipment to the system (Button-up perspective), enabling decision-making (top down) with the appropriate information of all assets and their affectation to the whole.

Overall, this study has successfully emphasized the importance of a clear management model and incorporating a systemic perspective in asset management and their digital twins. Embracing this vision and implementing the proposed framework will enable railways to harness the full potential of digital technologies and navigate the challenges and opportunities presented by the 4.0 industry.

Acknowledgements This paper has been written within the framework of the projects INMA "Asset Digitalization for INtelligent MAintenace" (Grant PY20 RE014 AICIA, founded by Junta de Andalucía PAIDI 2020, Andalucía FEDER 2014-2020) and Geminhi (Digital model for Intelligent Maintenance based on Hybrid prognostics models), (Grant US-1381456, founded by Junta de Andalucía, Andalucía FEDER 2014-2020).

References

Ali, K. N., et al. (2022). *Citation: Collaboration and risk in building information modelling (BIM): A systematic literature review*. https://doi.org/10.3390/buildings12050571

Bader, S., et al. (2018). *Details of the Asset Administration Shell Part 1 – The exchange of information between partners in the value chain of Industrie 4.0 (Version 3.0 RC02)* (p. 473). Federal Ministry for Economic Affairs and Climate Action (BMWK).

Candón, E., et al. (2019). Implementing Intelligent asset management systems (IAMS) within an industry 4.0 manufacturing environment. *IFAC-PapersOnLine, 52*(13), 2488–2493. https://doi.org/10.1016/J.IFACOL.2019.11.580

Carretero, J., et al. (2003). Applying RCM in large scale systems: A case study with railway networks. *Reliability Engineering and System Safety, 82*(3), 257–273. https://doi.org/10.1016/S0951-8320(03)00167-4

Crespo Márquez, A. (2007). *The maintenance management framework. Models and methods for complex systems maintenance*. Springer. ISBN 978-1-84628-821-0. https://doi.org/10.1007/978-1-84628-821-0

Crespo Márquez, A. (2022). Driving the introduction of digital technologies to enhance the maintenance management process and framework. In *Digital maintenance management: Guiding digital transformation in maintenance* (Springer Series in Reliability Engineering) (pp. 25–30). Springer International Publishing. https://doi.org/10.1007/978-3-030-97660-6_3

Leitner, B., Môcová, L., & Hromada, M. (2017). A new approach to identification of critical elements in railway infrastructure. *Procedia Engineering, 187*, 143–149. https://doi.org/10.1016/j.proeng.2017.04.360

Li, H., et al. (2022). Integrated representation of geospatial data, model, and knowledge for digital twin railway. *International Journal of Digital Earth, 15*(1), 1657–1675. https://doi.org/10.1080/17538947.2022.2127949

Macchi, M., et al. (2012). Maintenance management of railway infrastructures based on reliability analysis. *Reliability Engineering and System Safety, 104*, 71–83. https://doi.org/10.1016/j.ress.2012.03.017

Malakuti, S., et al. (2020). Digital twins for industrial applications. *IIC Journal of Innovation, February*, 1–19. Available at: https://www.iiconsortium.org/pdf/IIC_Digital_Twins_Industrial_Apps_White_Paper_2020-02-18.pdf

Mohammadi, A., & El-Diraby, T. (2021). Toward user-oriented asset management for urban railway systems. *Sustainable Cities and Society, 70*, 102903. https://doi.org/10.1016/J.SCS.2021.102903

RIH. (2023). *Railway Innovation Hub*. Available at: https://www.railwayinnovationhub.com/bim/. Accessed 23 July 2023.

Roda, I. (2022). *Maximing the benefits of big data for asset management*. UIC Rail System Department. ISBN 978-2-7461-3186-6.

Schweichhart, K. (2016). Reference architectural model industrie 4.0 (Rami 4.0). *An Introduction*. Available online: https://www.plattform-i40.deI40

Sedghi, M., et al. (2021). A taxonomy of railway track maintenance planning and scheduling: A review and research trends. *Reliability Engineering & System Safety, 215*, 107827. https://doi.org/10.1016/J.RESS.2021.107827

UIC A. W. G. (2022). 220430_UIC E2E decision tools landscape and high-level roadmap.pdf. Paris.

Weik, N., et al. (2022). DFT modeling approach for operational risk assessment of railway infrastructure. *International Journal on Software Tools for Technology Transfer, 24*, 331–350. https://doi.org/10.1007/s10009-022-00652-4

Zheng, X., Lu, J., & Kiritsis, D. (2021). The emergence of cognitive digital twin: Vision, challenges and opportunities. *International Journal of Production Research, 60*(2), 1–23. https://doi.org/10.1080/00207543.2021.2014591

Mauricio Rodríguez is a PhD candidate in Mechanical Engineering and Industrial Organization from the University of Seville with a solid background that includes a master's degree in Asset Management and Maintenance from the Federico Santa María Technical University, and a Postgraduate Degree in Transport Systems from the Polytechnic University of Madrid. He has extensive industrial and academic experience. Currently as part of the SIM Intelligent Systems Research group (of the University of Seville), he has contributed significantly to academic research and development by participating as a Senior Consultant in the INGEMAN Group, involved in projects focused on the improvement of industrial processes, Industry 4.0 and digital transformation, thus consolidating his integral role both in the academic and industrial fields achieving a effective two-way transfer.

Adolfo Crespo Márquez is currently Full Professor at the Department of Industrial Management, School of Engineering of the University of Seville. He holds a PhD with Honors in Industrial Engineering from this same university. He is included in the Stanford University's list of world top-scientists in 2023 and he received the Spanish Maintenance Award from the Spanish Maintenance Association (AEM) in 2020. With research works published in many top journals, he is the author of 8 books and editor of 6 books with Springer, Aenor, IGI Global and INGEMAN, about maintenance, warranty, supply chain and assets management. Prof. Crespo is Fellow Director of ISEAM, Editor-in-Chief of EAMR Springer Book Series and led the Spanish Research Network on Assets Management and the Spanish Committee for Maintenance Standardization (1995–2003). He also leads the SIM research group related to smart maintenance systems and has extensively participated in many engineering and consulting projects for different companies, for the Spanish Departments of Defense, Science and Education as well as for the European Commission (IPTS). He is the President of INGEMAN (a National Association for the Development of Maintenance Engineering in Spain) since 2002.

Antonio Jesús Guillén López obtained his PhD in Industrial Engineering from the University of Seville, currently holds the position of Professor in the Department of Organization of Companies at the University Complutense of Madrid. He is member of the SIM (Sistemas Inteligentes de Mantenimiento) research group at the University of Seville, contributing significantly to the field of intelligent maintenance systems. His career has been dedicated to impactful research, particularly in the areas of asset management and maintenance. His leadership extends to serving as the Secretary General of ESReDA since 2022, overseeing European Safety, Reliability, and Data Association initiatives. Since 2012, he has collaborated as a consultant with INGEMAN (Asociación para el Desarrollo de la Ingeniería de Mantenimiento) on projects with notable companies, including ADIF, ENAGAS, TALGO, AZVI, VIESGO, and ABENGOA.

Eduardo Candón Fernández is an Industrial Engineer from the School of Engineering of the University of Seville, member of INGEMAN (National Association for the Development of Maintenance Engineering in Spain) and member of the SIM (Intelligent Maintenance Systems) research group of the Department of Industrial Organization and Business Management of the School of Engineering of Seville, and currently a PhD student in the program of mechanical engineering and industrial management.

Determination of the Exact Economic Time for the Component Replacement Using Condition-Based Maintenance

Antonio Sánchez-Herguedas, Antonio Jesús Guillén-López, and Francisco Rodrigo-Muñoz

Abstract In most industrial assets, determining the preventive interval is a task carried out by the maintenance engineer. In non-critical assets, the optimization process of the interval must consider the costs of operation and maintenance, as well as the income generated by its operation. The result is the economic determination optimal moment to perform preventive intervention (PM). Mathematically, an expression can be found that relates these variables to the failure occurrence process. However, when the equipment is critical to the business, it is necessary to avoid the occurrence of failure. For this purpose, investment is made in techniques that determine asset degradation (CBM). In this case, not only must the failure occurrence process be controlled, but the degradation of the asset must also be analyzed. To determine the economically optimal moment for the preventive replacement of a component subject to CBM, a semi-Markovian model has been developed. The model considers degradation as a Wiener process and integrates it with the failure occurrence process, adjusted to a Weibull distribution. The result is two mathematical formulas to determine the optimal degradation threshold and the interval for preventive replacement, optimizing costs, income, degradation, and failure distribution.

Keywords Preventive interval · Income · Semi-Markovian model · Wiener process

---

A. Sánchez-Herguedas (✉) · A. J. Guillén-López
Department of Industrial Management, School of Engineering, University of Seville, Seville, Spain
e-mail: antoniosh@us.es; ajguillen@us.es

F. Rodrigo-Muñoz
Department of Applied Mathematics II, School of Engineering, University of Seville, Seville, Spain
e-mail: frodrigo@us.es

A. Crespo Márquez et al. (eds.), *Advances in Asset Management: Strategies, Technologies, and Industry Applications*, Engineering Asset Management Review 3, https://doi.org/10.1007/978-3-031-52391-5_7

1 Introduction

Many industrial assets are subject to condition-based maintenance (CBM) and post-failure corrective tasks. Once the degradation threshold value is reached, the preventive task is scheduled to avoid failure. The interval size to perform the preventive replacement is the key, i.e. how much time is running before the failure occurs? To calculate the preventive interval τ of a failure mode with degradation for an asset, it is necessary to carry out a study where costs, income, degradation, and probability of failure must be considered. It is usual for this study not to be carried out due to a lack of time, resources, or knowledge, and the asset owner or its maintainer chooses to set alarm thresholds obtained from their own experience or from the experience of others who are experts in the CBM of the asset. The asset that has been designed for the conditions set by the manufacturer is used by the owner in his particular conditions. The maintenance manager must strive to calculate the economically optimal degradation threshold and preventive interval under his conditions of use and maintenance. By using degradation levels, CBM policies reduce the occurrence of faults and lengthen the periods between preventive tasks (Table 1).

A semi-Markovian model is designed to observe the influence on the size of the optimal preventive interval of the asset's uptime income, the costs of corrective and preventive maintenance interventions, degradation, and failures. This model presents a state where the corrective task is performed, a preventive state, and two operational states. The first of these two states is controlled by item degradation. The degradation is defined by a mathematical function following a Wiener process. In this state, the effective degradation of the element is measured by CBM techniques. The probability of asset failure is considered zero as long as the degradation does not exceed a set threshold. Once this threshold is exceeded, the asset is placed in the second operational state. From this point on, the probability of failure is no longer zero and the model is controlled by a Weibull distribution.

The model evolves in time, following a semi-Markovian process with an embedded Markov chain. The process is semi-Markovian because the sojourn times in the states are not exponential functions. Transitions between states occur according to

Table 1 List of terms/nomenclature

CBM	condition-based maintenance	τ_o	optimal preventive interval
PHM	Prognostics and health management	RUL	Remaining useful life
R_i	Cost or income in state i	R_{ij}	Cost of transition from i to j
D_o	Degradation threshold	T_o	Degradation threshold time
V(m)	Expected accumulated return in m transitions	V(1)	Expected accumulated return in one transition
P	Transitions matrix	R	Returns matrix
τ	Preventive interval	D(t)	Degradation function
W(t)	Wiener process	σ	Wiener process drift
ν	Wiener process mean trend	α	Weibull sharpe parameter
β	Weibull scale parameter		

probabilities fixed in the Markov chain. At each transition, the process accumulates costs or income as returns (negative or positive). The aim is to find the preventive interval that maximizes the accumulated returns at each transition. This allows finding optimal values of the preventive interval τ_o for any transition of the time horizon. The model proposes a system of difference equations for expected accumulated return, solved by applying the z-transform. Subsequently, by derivation the mathematical expression of the preventive interval τ_0 is reached.

The model allows us to analyze how the preventive interval is affected by changes in income, costs, degradation, and failures. In addition, it allows the calculation of the threshold limit of degradation for the signal collected in the monitoring for this optimal preventive interval. These two decisions are the key to establishing a CBM policy. This last value is so far calculated based on experimental values and the expert's experience. This work is included among works issues related to Prognostics and Health Management (PHM).

2 Background

The calculation of the value of the optimal preventive interval is analyzed by many authors, who have used different techniques and models to obtain it. Some of them have elaborated models built based on semi-Markovian processes. This technique is executed transition by transition, visiting successive states, so it replicates well the sequence of execution of operation and maintenance activities of industrial physical assets.

Hu et al. (2020) use a Markov model with preventive substitutions and imperfect repairs. In the paper, they develop a semi-Markovian model to find the value of the optimal preventive interval, minimizing the long-term average cost. As in our work, the result is a mathematical formula valid for any time horizon. Other authors also use this technique. Lyubchenko et al. (2018) present an application of this approach to evaluate the recommended preventive maintenance intervals in radio devices. In this case, they use Markov chain theory to mathematically describe the sequence of transitions between states and apply the semi-Markovian process model to the random process of sojourn times in each state. Wang and Miao (2021) formulate a preventive maintenance optimization model under a semi-Markovian model for a balanced system, where each unit is subject to degradation failure and the dwell times in each state follow Erlang distributions with different parameters. Grabski (2014) analyses the technique and explains how to build semi-Markovian models, discussing the different parameters and reliability characteristics that can be obtained from these models. He defines the properties and theorems of the theory of semi-Markovian processes.

Other authors combine semi-Markovian models with other techniques and theories. To develop their model, Kumar and Varghese (2018) use non-exponential failure and repair time distributions. This forces them to model from a semi-Markovian approach. But they focus their attention on the evaluation of system availability.

They then derive the availability-optimizing preventive interval using the golden section search technique. Wu et al. (2021) develop their model using continuous-time semi-Markovian processes and then provide a method for solving these processes. They use algorithms from discrete-time cases, considering the error made when discretizing. This method is applied to a reliability problem, analyzing the availability of a system subject to sequential cyber-attacks. The method solves two scenarios when the sojourn times follow exponential or Weibull distributions. Kumar et al. (2021) propose a semi-Markovian approach to analyze the degradation of complex mechanical systems by constructing operational states. They start from an initial scenario and consider the failure rate and the repair rate to establish the transition probabilities between states. Finally, they calculate the system availability when the stationary scenario is reached. They perform a practical application of this proposed methodology to analyze the degradation of an air compressor and calculate its availability. Wu et al. (2019) propose a competitive risk model with a constraint for transition times in repairable multistate systems following semi-Markovian processes. Once the model is established, they employ aggregate stochastic processes to obtain the formulas for the competitive risk probabilities, survival time distributions, and availabilities. Nasrfard et al. (2022) propose a probabilistic approach considering some correlations and uncertainties to find the optimal inspection rates. They develop a model using a semi-Markovian chain based on Monte Carlo simulations, using 95% percentiles of the total cost to determine the optimal inspection rates. Wang et al. (2019) consider a condition- and age-based optimal maintenance policy for a repairable two-unit serial system. They formulate and solve the maintenance problem in the semi-Markovian decision process framework. They establish a formula for the average maintenance cost and determine the optimal levels for the maintenance of the two units that minimize the long-term average cost.

Other authors use Markov chains. Farahani et al. (2019) model a production system as a continuous-time Markov chain. The model determines the optimal preventive maintenance interval by reducing the unit time costs of corrective and preventive interventions. Farhadi et al. (2022) also use Markov chains, in this case, to determine the optimal number of spare parts and establish the best supplier and the appropriate quality of the spare part. They suggest several models and apply numerical samples to show the state representation and determine optimal spare parts supply strategies and inventory policies.

Our model includes the income per unit of asset operating time after degradation in the preventive interval calculation. This revenue received by the owner is not considered in most optimization studies. There are very few cases where this data is included in the analysis, and this work aims to fill this gap as well. Zhu et al. (2021) develop a preventive maintenance optimization model based on a three-stage failure process for a single-component system. The objective is to maximize profit, but unlike conventional optimization models, it uses a revenue function to correlate profit with availability and cost. Mizutani and Zhao (2021) use various reliability engineering techniques and tools to choose the best strategies for systems with replacements, including periodic replacements. The developed method searches for the optimal preventive interval. To do so, they start from the survival function and the values of revenues and costs during operation and maintenance interventions. Our study is a continuation of other articles published using the income obtained from the operation of assets. (Sánchez-Herguedas et al., 2021, 2022, 2022a, b, and c) for the optimal preventive interval calculation.

In the literature, it is rare to see the use of income in optimization. Formulas to calculate the preventive interval and the degradation threshold for any transition are also not usually given. The degradation threshold is usually calculated in relation to the failure. In this case, it is calculated in relation to the time of preventive replacement.

Degradation data have been widely used to predict the remaining lifetime of systems. Most previous work uses a preset model to capture the degradation process and focuses on degradation processes without constant shocks or shock effects. (Kong et al., 2021). Stogiannis and Caroni (2013) use a time-to-first impact model based on a Wiener process to determine the degradation of a component, furthermore, they attempt to fit this model to data generated by a Weibull regression. Liu et al. (2017) review the developments of Wiener process-based models for degradation data analysis and RUL estimation, as well as their applications in forecasting and asset health management. In addition, they discuss applications of Wiener process-based models for degradation test design and optimal decision-making activities, such as inspection, condition-based maintenance, and replacement. In the end, they highlight several future challenges that deserve further study. However, in our case, a degraded operating state is included. The word degraded is intended to express the circumstance that the failure mode of the element follows the same failure distribution function from which it started in the operational state, even though the income generated by its operation is lower in this second state. However, mathematical development is also valid whether the higher income. That opens the door to analyzing other problem types.

The main contribution of this study is the development of a mathematical formula that calculates the optimal preventive interval when degradation is measured quantitatively. This formula includes the costs of maintenance interventions performed when condition-based maintenance is applied: corrective interventions to restore the asset in case of failure and preventive interventions to reduce the failure probability due to wear and tear. It also includes the asset income when operating in degraded conditions, from which the costs derived from the degradation generated signal can be subtracted. The formula can be applied to a time horizon without limitation or to cases where the use of the asset is limited, for example, by the completion of the business project. This second case is the most common in a critical asset subject to CBM. The developed formula is easy to apply for the designer of asset maintenance plans and can also be included in the designs of digital twins of assets and maintenance processes.

3 Material and Methods

The first objective is to find the mathematical expression of the preventive interval τ_0 that optimizes the expected accumulated return over time, transition by transition. The second objective is to find the threshold limit of degradation D_0. Both magnitudes depend on the costs of the maintenance tasks, the operating income, possible penalties for their inactivity, the evolution of the degradation, and the distribution of their failures. In this section, the characteristics of the mathematical model are presented. In the following section, the proposed model is solved mathematically. The result is two mathematical formulae that a maintenance manager can apply to his plant assets without the knowledge of modeling and calculation techniques. In this section, the semi-Markovian model is designed and developed.

3.1 *Model, Returns, Degradation and Failure Distribution*

A four-state model is developed and could be applied to the case of the failure of a turbine shaft bearing. The first state S_1 corresponds to the situation where the asset is operational, its owner receives an income R_1 for its use, but the element under study suffers degradation. This degradation can be controlled by some predictive technique (in this case by vibration analysis). The asset remains in S_1 until the vibration value first reaches the degradation threshold value D_0. This value as far as we know is a value calculated experimentally by the technicians and is not based on mathematical calculations but on their own experience. The transition to state S_4 can incorporate a cost of R_{14} into the model. To calculate the economically optimal value of D_0 we must, on the one hand, estimate the average time T_0 until D_0 is first reached. On the other hand, it must be assumed that the probability of bearing failure up to D_0 is zero. This assumption is the basis for the application of any predictive maintenance technique. To calculate the value of T_0 a degradation function integrating a Wiener process is considered.

Once degradation D_0, is reached, the asset enters the degraded state S_4. As the degradation continues, the probability of failure increases. It can be represented using a failure distribution function. In our case, we fit the experimental function from the censored and failure data to a three-parameter Weibull distribution function. Operation in this state also produces income R_4, which may be different from that of the S_1 state. A preventive intervention must be scheduled in this state because failure is near. If the bearing fails, which it does with a probability described by the Weibull function, the equipment reaches the corrective state S_2. If a preventive task is performed before the failure, the equipment reaches the preventive state S_3.

If the equipment fails, the system incurs costs due to transition R_{42}. In state S_2, the system incurs costs of types R_2 and R_{21}. If after a time τ of operation from the initial instant, the equipment undergoes the preventive intervention. The system incurs costs R_{43} due to the transition. In state S_3, the system incurs costs of types R_3

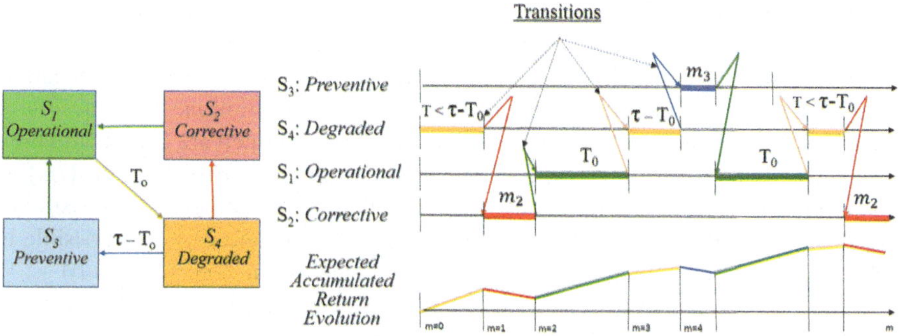

Fig. 1 Representation of states, transitions between states, and expected accumulated returns at each transition

and R_{31}. Because of remaining in each state, costs and incomes (R_i and R_{ij}) are accumulated over time in a variable called the expected accumulated return. See Fig. 1. The model objective is to find the value of τ_0 and D_0 (T_0) that optimizes the expected accumulated return.

The returns (incomes and costs) involved in the process are:

- R_1, income per unit time that the system remains in state 1 (S_1: Operational).
- R_2, cost per unit time that the system remains on state 2 (S_2: Corrective)
- R_3, cost per unit time that the system remains on state 3 (S_3: Preventive).
- R_4, income per unit time that the system remains in state 4 (S_4: Deteriorated).
- R_{14}, transition cost from operational state to degraded.
- R_{21}, transition cost from corrective state to operational.
- R_{31}, transition cost from preventive state to operational.
- R_{42}, transition cost from degraded state to corrective.
- R_{43}, transition cost from degraded state to preventive.

3.2 Semi-Markovian Maintenance Model

In this work, a Semi-Markov Process (SMP) is selected as the stochastic format of the maintenance model. These processes are powerful tools for reliability optimization and maintenance problems solver (Hu & Yue, 2003; Kim & Makis, 2009, 2010; Zhang & Gao, 2012). Those are utilized to model the impact of maintenance strategies in a system and for a finite number of periods.

The designed semi-Markovian model consists of two items, a semi-Markovian process with transitions between states and sojourn times and a homogeneous Markov chain embedded in the process. The process defines the state space, in the general case $\{X_n, n \geq 0\}$ with n states. The chain defines the state transition probabilities at each transition $p_{ij} = P(X_1 = j | X_0 = i)$. This chain determines the evolution of the process. The permanence time of the state and the length of the time horizon are determined by the process. The semi-Markovian process is characterized by the fact that sojourn times differ in each state. The sojourn times and the transition between states have associated economic returns that can be positive or negative. The variable $r_{ij}(m)$ contains the return from i to j in transition m. In m successive transitions from state i, the process accumulates returns, which are added with their respective signs, constituting the accumulated return in m transitions. Let $R_i(m)$ be this random variable. The model can establish many alternatives in m transitions. For this reason, the value of $R_i(m)$. is impossible to calculate. But it is possible to calculate its average value, the expected accumulated return $v_i(m) = E(R_i(m))$.

Following the process established by (Sánchez Herguedas et al., 2022) the following system of differential equations for the mean cumulative return, expressed in vector form, is reached:

$$V(m) = V(1) + P \cdot V(m-1) \tag{1}$$

Degradation Process

The asset operating time is a random variable T and can take any positive value. Usually, this variable is called time to failure. If the system is in operational state S_1, it will remain in this state until degradation reaches a threshold D_0, for the first time, moving to the degraded state, S_4. T_0 is the average time the system remains in the S_1 before the degradation D_0 is reached for the first time. We will assume that before reaching the threshold D_0 the system has such a low failure rate that we can assume zero probability of failure.

From its start-up (state S_1), the item whose failure mode is being analyzed is subjected to a degradation process that increases with time, starting from an initial value of zero. This process is modeled (Letot et al., 2015) by a linear Gaussian degradation process, as expressed in the equation:

$$D(t) = D(0) + vt + \sigma W(t) \tag{2}$$

The parameter v represents the mean trend and σ the drift. W(t) is a Wiener process characterized by having a zero initial value, $W(t) = 0$, by being a continuous function concerning time, and by having independent transitions that can be expressed by a Normal according to $W(t) - W(s) = N(0, t - s)$. The increments have Normal distribution.

From this, it follows:

$$W(t) = W(t) - 0 = W(t) - W(0) \sim \mathcal{N}(0, t-0) = \mathcal{N}(0, t),$$

That is, for each $t \geq 0$, the random variable $W(t)$ has a normal distribution of mean 0 and variance t.

Being a continuous stochastic process, the function $W(t)$ is continuous.

In the Wiener process, two interesting equalities will be used in the practical application of the treatment of the degradation data:

(a) The random variables $W(T\ .\ t)$ and $\sqrt{T}\cdot W(t)$ where t, $T \geq 0$ have the same probability distribution.
(b) The random variables $W(t)$ and $\sqrt{t}\cdot Z$ where $t \geq 0$ y $Z \sim \mathcal{N}(0,1)$, have the same probability distribution.

The degradation D_0 is reached at the instant:

$$T_0 = \inf\{t : D(t) \geq D_0\} = D^{-1}(D_0) \tag{3}$$

The use of a Wiener process implies that T_0 is a random variable with an inverse Gaussian distribution that has a probability density function (Wang, 2010):

$$f_0(t) = \frac{D_0}{\sigma\sqrt{2\pi t^3}} \exp\left\{-\frac{(D_0 - vt)^2}{2\sigma^2 t}\right\}, v > 0,\ t > 0 \tag{4}$$

Sojourn Time in Each State

But when the system reaches the degraded state, the level of degradation has exceeded the threshold D_0. The probability of failure can no longer be assumed to be zero; from this, the system may fail. To prevent the failure, after a non-random time τ and knowing that $\tau > T_0$, the system undergoes a preventive intervention that returns it to the initial situation. However, the failure may occur before the instant τ, in which case the system must undergo a corrective intervention to return it to the initial situation.

According to this mechanism, the system remains in the state S_4 from the instant T_0 (measured from the initial instant), when the degradation threshold was reached, until a failure occurs or until the instant τ (from the initial instant), is reached, depending on which of the two occurs first. The instant of exit from S_4 is therefore $\min\{T, \tau\}$, so the sojourn time in S_4 is:

$$T_4 = \min\{T, \tau\} - T_0 = \min\{T - T_0, \tau - T_0\} \tag{5}$$

Where T is the time-to-failure random variable. Since the system remains in the state S_4 between T_0 and at most τ, the sojourn time T_4 is the random variable $T - T_0$ truncated to the interval $[T_0, \tau]$. If the failure occurs, it will undergo repair and return to state S_1. The time taken for the repair is also a random variable T_2. If it does not fail preventive task occurs and later transits to state S_1. The time taken for the preventive intervention is also a random variable T_3.

This maintenance model can be seen as a stochastic process with a space of four states $\{S_1, S_2, S_3, S_4\}$. The times the system remains in each state are random variables.

The distribution functions of T, T_2, T_3 we will designate $F(.)$, $F_2(.)$, $F_3(.)$ and the probability density functions $f(.)$, $f_2(.)$, $f_3(.)$. The random variable T_4 truncated distribution whose distribution function is defined from the distribution function of T.

The matrix of sojourn times in one state before moving on to another is:

$$\begin{pmatrix} 0 & 0 & 0 & T_0 \\ T_2 & 0 & 0 & 0 \\ T_3 & 0 & 0 & 0 \\ 0 & T_4 & \tau - T_0 & 0 \end{pmatrix} \tag{6}$$

The average sojourn times in the operational, corrective, and preventive states are the averages of the random variables T_0, T_2 and T_3:

$$A = E(T) = \int_{\infty}^{0} tf_1(t)dt, \; B = E(T_2) = \int_{\infty}^{0} tf_2(t)dt = m_2,$$

$$C = E(T_3) = \int_{\infty}^{0} tf_3(t)dt = m_3 \tag{7}$$

The mean of T_4, i.e. the average sojourn time in the degraded state is:

$$D = E(T_4) = \frac{1}{F(\tau) - F(T_0)} \int_{\tau}^{T_0} (t - T_0) f(t) dt = \frac{1}{F(\tau) - F(T_0)} \int_{\tau}^{T_0} tf(t)dt - T_0$$

Since $F(T_0) = 0$, the expressions of D are transformed into:

$$D = E(T_4) = \frac{1}{F(\tau)} \int_{\tau}^{T_0} (t - T_0) f(t) dt = \frac{1}{F(\tau)} \int_{\tau}^{T_0} tf(t)dt - T_0 \tag{8}$$

The average sojourn times matrix is the average of matrix sojourn times (Eq. 6):

$$Q = E\left[\begin{pmatrix} 0 & 0 & 0 & T_0 \\ T_2 & 0 & 0 & 0 \\ T_3 & 0 & 0 & 0 \\ 0 & T_4 & \tau - T_0 & 0 \end{pmatrix}\right] = \begin{pmatrix} 0 & 0 & 0 & A \\ B & 0 & 0 & 0 \\ C & 0 & 0 & 0 \\ 0 & D & \tau - T_0 & 0 \end{pmatrix} \tag{9}$$

Where $E[.]$ means, mean value.

Transition Probabilities Between States

As in all continuous-time random processes, while our semi-Markovian process is in one state, there is no moment when it transitions to the same state it is already in. It remains there until the next transition takes it to a different state. The probability of transition from a state to itself is zero. Being in state S_1, the system can only transit to state S_4, so the transition probability is $p_{14} = 1$, and it cannot transit to the other states, so $p_{11} = p_{13} = p_{14} = 0$.

While the system is in state S_4, it can only move to states S_2 and S_3. The transition to state S_2 will occur if the time to failure T is longer than T_0 (indicating that it is no longer in state S_1) and less than τ. Thus, the probability of transit to state S_2 is:

$$p_{42} = P\left(T \langle \tau \mid T > T_0\right) = \frac{P\left[\left(T < \tau\right) \cap \left(T > T_0\right)\right]}{P\left(T > T_0\right)} = \frac{P\left(T_0 < T < \tau\right)}{1 - P\left(T < T_0\right)}$$
$$= \frac{F\left(\tau\right) - F\left(T_0\right)}{1 - F\left(T_0\right)} = F\left(\tau\right) = p$$

Since $F(T_0) = 0$. The probability of transiting to state S_3 is the complementary one:

$$p_{43} = P\left(T > \tau \mid T \rangle T_0\right) = 1 - P\left(T \langle \tau \mid T > T_0\right) = 1 - p.$$

From states S_2 and S_3, the system can only move to state S_1, so the transition probabilities p_{21} and p_{31} are both 1.

According to these considerations, the transition probability matrix is:

$$P = \begin{pmatrix} p_{11} & p_{12} & p_{13} & p_{14} \\ p_{21} & p_{22} & p_{23} & p_{24} \\ p_{31} & p_{32} & p_{33} & p_{34} \\ p_{41} & p_{42} & p_{43} & p_{44} \end{pmatrix} = \begin{pmatrix} 0 & 0 & 0 & 1 \\ 1 & 0 & 0 & 0 \\ 1 & 0 & 0 & 0 \\ 0 & p & 1-p & 0 \end{pmatrix} \tag{10}$$

This is a stochastic matrix because it is the probability matrix of a Markov chain. The transitions of this chain occur every time the stochastic process changes state, so it is the Markov chain embedded in the process.

We have developed a stochastic process that models the maintenance of our system, which has a finite state space, the sojourn times in each state are random variables with probability distributions not necessarily exponential, and has an embedded Markov chain whose transition probabilities matrix is Eq. 10. We conclude that this is a semi-Markovian process.

Returns Matrix

The matrix of returns is the matrix whose $ij-th$ element is the sum of the return due to remaining in the $i-$ th estado state and that due to the transition to the $j-$ th state:

$$R=\begin{pmatrix} 0 & 0 & 0 & AR_1+R_{14} \\ BR_2+R_{21} & 0 & 0 & 0 \\ CR_3+R_{31} & 0 & 0 & 0 \\ 0 & DR_4+R_{42} & (\tau-T_0)R_4+R_{43} & 0 \end{pmatrix} \tag{11}$$

4 Development of Mathematical Expressions

4.1 Expected Accumulated Return

In order to solve the Eq. 1, the z-transform will now be used. Remember that the z-transform of a sequence $x(m)$, which is usually denoted $\mathcal{Z}[x(m)]$, is a complex variable function defined as the Laurent series:

$$\mathcal{Z}[x(m)]=\sum_{\infty}^{n=0} x(n)z^{-n},\ z\in C\setminus\{0\}. \tag{12}$$

A vector whose components are sequences is considered. The z-transform of a vector is another vector whose components are the z-transforms of the components. For convenience, the Eq. 1 is rewritten with the index increased by one unit:

$$V(m+1)=V(1)+PV(m). \tag{13}$$

Multiplying by the inverse of the regular matrix $I-z^{-1}P$, dividing by z and reordering terms, the Eq. 14 is obtained:

$$\mathcal{Z}[V(m)]=\frac{1}{z-1}\left(I-z^{-1}P\right)^{-1}V(1)+\left(I-z^{-1}P\right)^{-1}V(0) \tag{14}$$

The initial expected returns vector $V(0)$ is the null vector since there is no return before the initial instant when the chain evolution begins. During the development, we need the vector $V(1)$, which is obtained from $v_i(1)=\sum_{4}^{j=1} r_{ij}(1)p_{ij}\ with\, i=1,2,3,4$ for which we need the Probability P and Returns R matrices:

$$V(1)=\begin{pmatrix} v_1(1) \\ v_2(1) \\ v_3(1) \\ v_4(1) \end{pmatrix}=\begin{pmatrix} AR_1+R_{14} \\ BR_2+R_{21} \\ CR_3+R_{31} \\ p(DR_4+R_{42})+(1-p)((\tau-T_0)R_4+R_{43}) \end{pmatrix} \tag{15}$$

To solve the Eq. 13, it remains is to invert these z-transforms. This process can be followed in (Sánchez Herguedas et al., 2022).

After a complex development in z-transforms, where initially the rational functions in z are considered:

$$\frac{z^3}{(z-1)(z^3-1)}, \frac{z^2}{(z-1)(z^3-1)}, \frac{z}{(z-1)(z^3-1)}$$

And considering the circumstances that the variable z only appears in simple fractions:

$$\frac{1}{z-\left(-\frac{1}{2}-\frac{\sqrt{3}}{2}\mathrm{i}\right)}, \frac{1}{z-\left(-\frac{1}{2}+\frac{\sqrt{3}}{2}\mathrm{i}\right)}, \frac{1}{(z-1)^2}, \frac{1}{z-1},$$

We develop $\mathcal{Z}[V(m)]$ in terms of these four fractions and call their coefficients B_1, B_2, A_3 and A_4. B_1 and B_2 are complex conjugates.

After applying Laurent's series of the four simple fractions and for the sake of simplification, we call r and θ the modulus and argument of $-\frac{1}{2}-\frac{\sqrt{3}}{2}\mathrm{i}$, that is:

$$r=\sqrt{\left(-\frac{1}{2}\right)^2+\left(-\frac{\sqrt{3}}{2}\right)^2}=1,\ \theta=\arg\left(-\frac{1}{2}-\frac{\sqrt{3}}{2}\mathrm{i}\right)=\arctan\left(\frac{-\frac{\sqrt{3}}{2}}{-\frac{1}{2}}\right)+\pi=\frac{4\pi}{3}$$

The inverse transformation of $\mathcal{Z}[v_1(m)]$, that is $v_1(m)$, the first component of the expected accumulated return is as follows:

$$v_1(0)=0 \tag{16}$$

$$v_1(m)=2\left(Re(B_1)\cos\frac{4(m-1)\pi}{3}-Im(B_1)\sin\frac{4(m-1)\pi}{3}\right)$$
$$+A_3(m-1)+A_4,\ m=1,2,3,\cdots$$

$$B_1 = \frac{1}{6}\left[v_1(1) - v_4(1) + \frac{1}{\sqrt{3}} i\left(v_1(1) + v_4(1) - 2p\left(v_2(1) - v_3(1) \right) - 2v_3(1) \right) \right]$$

$$A_3 = \frac{1}{3}\left[v_1(1) + v_4(1) + p\left(v_2(1) - v_3(1) \right) + v_3(1) \right]$$

$$A_4 = \frac{1}{3}\left[2v_1(1) + v_4(1) \right]$$

For each transition m, a value of the expected accumulated return is obtained.

4.2 Optimal Preventive Interval

Let's introduce the failure distribution function. A Weibull distribution function with shape parameter α and scale parameter β. The assumption that there are no failures before the degradation threshold D_0 is reached implies a third parameter T_0 that represents the average sojourn time of the system in the operational state, guaranteed life. Then the probability density functions $f(t)$ and distribution $F(t)$ are:

$$f(t) = \frac{\alpha}{\beta}\left(\frac{t - T_0}{\beta} \right)^{\alpha - 1} e^{-\left(\frac{t - T_0}{\beta} \right)^{\alpha}}, \quad F(t) = 1 - e^{-\left(\frac{t - T_0}{\beta} \right)^{\alpha}}, \quad \alpha, \beta > 0$$

By substituting them in Eq. 16, and deriving with respect to the preventive interval τ, the expression is reached:

$$\frac{dv_1(m)}{d\tau} = \frac{1}{3}\left[\begin{array}{l} \frac{\alpha}{\beta}\left(\frac{\tau - T_0}{\beta} \right)^{\alpha - 1} e^{-\left(\frac{\tau - T_0}{\beta} \right)^{\alpha}} \left[\begin{array}{l} \left(R_{42} - R_{43} \right)\left(m + \frac{2}{\sqrt{3}} \sin\frac{4m\pi}{3} \right) \\ + \left(BR_2 + R_{21} - CR_3 - R_{31} \right) \\ \left(m - 1 - \frac{1}{\sqrt{3}} \sin\frac{4m\pi}{3} + \cos\frac{4m\pi}{3} \right) \end{array} \right] \\ + e^{-\left(\frac{\tau - T_0}{\beta} \right)^{\alpha}} R_4 \left(m + \frac{2}{\sqrt{3}} \sin\frac{4m\pi}{3} \right) \end{array} \right] \quad (17)$$

Equating to zero, we obtain Eq. 18 hat calculates the value of the optimal preventive interval for each transition m.

$$\tau_0 = T_0 + \left(\frac{\beta^\alpha}{\alpha} \cdot \frac{-R_4}{R_{42} - R_{43} + \left(BR_2 + R_{21} - CR_3 - R_{31}\right) \dfrac{m - 1 - \frac{1}{\sqrt{3}} \sin\frac{4m\pi}{3} + \cos\frac{4m\pi}{3}}{m + \frac{2}{\sqrt{3}} \sin\frac{4m\pi}{3}}} \right)^{\frac{1}{\alpha - 1}} \quad (18)$$

5 Degradation Threshold Calculation Process

The process of calculating the degradation threshold is an iterative process that starts from the choice of an experimental value for T_0. To calculate the optimal preventive interval value for a given transition (value of m), Eq. 18 is applied. With the values obtained for the experimental value of T_0 and the value obtained for τ_0, the expected accumulated return value $v_1(m)$ is calculated for the transition m chosen according to the duration of the business project. From these three values, the iterative process begins.

The first action consists of verifying the value of $v_1(m)$ when the value of T_0 is increased or decreased by a given amount. We call T_0^1 the value with the increase or decrease over T_0 that improves the value of $v_1(m)$ up to the value $v_1(m)^1$. With this new value of T_0^1, the new value τ_0^1 is calculated to obtain the new couple $\left(T_0^1, \tau_0^1\right)$. The iteration process continues until the values between the components of the last two couples $\left(T_0^n - T_0^{n-1}\right)$ and $\left(\tau_0^n - \tau_0^{n-1}\right)$ are lower than a previously established value that depends on the failure mode studied. In this case, the duplicate obtained from the n iteration would be the couple $\left(T_0^*, \tau_0^*\right)$ that maximizes the value $v_1(m)^n$.

For each data set from returns, degradation, and failures, an optimal couple (T_0^*, τ_0^*) for each transition m is calculated. An example of the iterative calculation process of the optimal couple is shown in Fig. 2.

The T_0^* value will determine the degradation threshold D_0^* that the technician must seek by applying predictive techniques. For this, Eq. 3 is used.

6 Results and Discussion

A mathematical model has been designed, developed, and solved in previous sections. Four of the fundamental states in which the assets find themselves during the operation and maintenance phase have been represented in the model. This model has two distinct parts. The first corresponds to the inspection stage in CBM, where inspections are carried out to obtain the value of the degradation signal. The stage

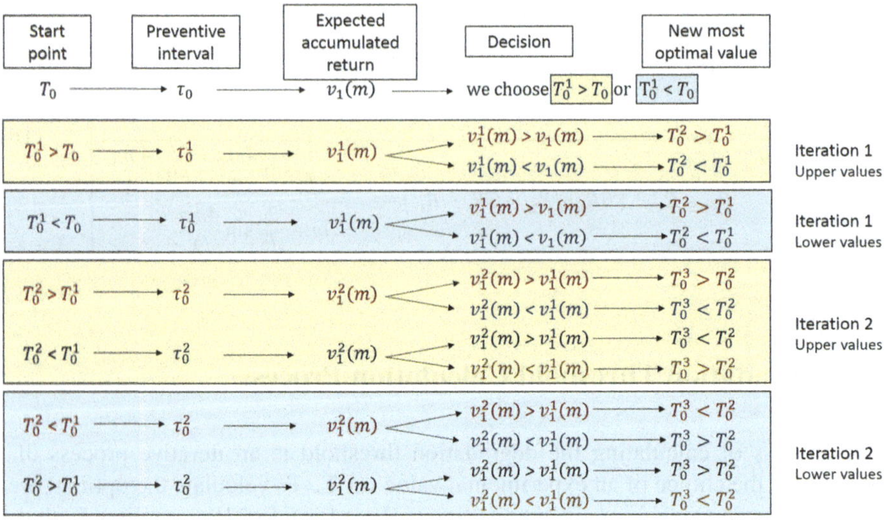

Fig. 2 Calculation of the couple (T_0^*, τ_0^*) that maximizes $v_1(m)$. Two iterations

objective is to locate the optimal degradation threshold, after which the preventive activity is decided. The second part corresponds to the optimal interval search, after which the preventive activity is performed.

From this model, two mathematical formulae are developed that will be used to obtain two-time values. The first formula corresponds to the degradation inspection stage. In the model, the degradation process is identified as a Wiener process. During its development, the mathematical formula that determines the time to reach a selected degradation for the first time is obtained. The linear fitting of the degradation of the asset element to the Wiener process requires the choice of two parameters, the linear trend, and the drift. For some cases, the linear trend must be substituted because not all elements show a linear degradation. For these types of assets, it must be replaced by other trend types. The second formula is developed from the semi-Markovian model. The economic optimization of the location of the preventive intervention is sought. This formula has been verified using the Nelder-Mead numerical optimization method and by continuous simulation models.

The model requires the interaction of these two formulas. This interaction generates a pair formed by two-time values. When the alert is established (degradation threshold) and when the preventive intervention is executed. Maintenance technicians must consider both times in the CBM activities management. The last step of the model is to economically optimize both times, i.e. to find the optimal couple. Possibly the couple will not be formed by the two optimal times. In this case, the expected accumulated return, the same variable used for the optimization of the preventive interval, is used as the optimization variable. From here, the degradation threshold can be determined. Its calculation comes from the degradation formula.

The model has been designed to support the information needs. The maintenance plan manager must consider them when making decisions. For this purpose, three relevant aspects have been considered. Those resemble the model of the maintenance process and condition the calculation scenario. Firstly, it calculates the optimum degradation level and preventive interval. Second, the calculation considers the costs of operation and maintenance activities and the income obtains from its operation by the owner. Third, it develops all the formulas to establish the optimal result for any operating horizon, as this selected horizon must coincide with the duration of the industrial project.

7 Conclusions

When applying CBM, several decisions must be made. The first decision is to determine the technique to capture the degradation signal. The second is to establish the signal threshold from which the probability of failure is not zero. The third is to determine the economically convenient interval to carry out the preventive intervention before the failure occurs. This article presents a tool composed of two mathematical formulas that respond to the last two decisions. To find the optimal preventative interval and degradation threshold when applying a CBM policy.

The formula of the preventive interval depends directly on the income in the degraded state (R_4), on the costs before reaching other states (R_{42} and R_{43}), and on the costs associated with corrective state S2 and preventive S3. It also depends on the component failure probability (α, β) and the time T_0 until the degradation threshold D_0 is reached.

Looking at the tools, the following observations are made:

- An increase of R_4 will lead to an increase in the optimal preventive interval.
- An increase in the difference between corrective and preventive costs increases, and the optimal preventive interval decreases.
- Once the threshold limit of degradation has been reached for the first time, the asset can continue to be used, although preventive intervention should be planned for the time marked by the optimal preventive interval.
- Although the interval formula allows one to calculate its value for any transition m, it is necessary to determine the transition that coincides with the project duration in which the asset is involved (finite horizon).

The limitation of this model, like most mathematical models, is that the input data for the calculations are obtained from failure events that have occurred in the past. Failures occur with some randomness but contain a bias due to wear that makes them more predictable. When this wear bias is significant, these models are useful. However, their use would be limited in situations where the randomness of the failure is not negligible. On the other hand, the formulas can be used in any digitization process at the failure mode level.

References

Farahani, A., Tohidi, H., & Shoja, A. (2019). An integrated optimization of quality control chart parameters and preventive maintenance using Markov chain. *Advances in Production Engineering And Management, 14*(1), 5–14. https://doi.org/10.14743/apem2019.1.307

Farhadi, M., Shahrokhi, M., & Rahmati, S. H. A. (2022). Developing a supplier selection model based on Markov chain and probability tree for a k-out-of-N system with different quality of spare parts. *Reliability Engineering and System Safety*. Elsevier Ltd, *222*(108387), 108387. https://doi.org/10.1016/j.ress.2022.108387

Grabski, F. (2014). *Semi-Markov processes: Applications in system reliability and maintenance*. Elsevier. https://doi.org/10.1016/C2013-0-14260-2

Hu, Q., & Yue, W. (2003). Optimal replacement of a system according to a semi-Markov decision process in a semi-Markov environment. *Optimization Methods and Software, 18*(2), 181–196. https://doi.org/10.1080/1055678031000111803

Hu, J., Shen, J., & Shen, L. (2020). Periodic preventive maintenance planning for systems working under a Markovian operating condition. *Computers and Industrial Engineering*. Elsevier, *142*(January), 106291. https://doi.org/10.1016/j.cie.2020.106291

Kim, M. J., & Makis, V. (2009). Optimal maintenance policy for a multi-state deteriorating system with two types of failures under general repair. *Computers and Industrial Engineering, 57*(1), 298–303. https://doi.org/10.1016/j.cie.2008.11.023

Kim, M. J., & Makis, V. (2010). A maintenance model with minimal and general repair. *IMA Journal of Management Mathematics, 21*, 385–393. https://doi.org/10.1093/imaman/dpn016

Kong, X., Yang, J., & Li, L. (2021). Remaining useful life prediction for degrading systems with random shocks considering measurement uncertainty, *Journal of Manufacturing Systems*. Elsevier Ltd, 61(December 2020), pp. 782–798. https://doi.org/10.1016/j.jmsy.2021.05.019

Kumar, G., & Varghese, J. P. (2018). Optimum preventive maintenance policy for a mechanical system using semi-Markov method and Golden section technique. In *2018 IEEE international conference on Industrial Engineering and Engineering Management (IEEM)*. pp. 232–236. https://doi.org/10.1109/IEEM.2018.8607343

Kumar, V., et al. (2021). Degrading systems availability analysis: Analytical semi-markov approach. *Eksploatacja i Niezawodnosc, 23*(1), 195–208. https://doi.org/10.17531/EIN.2021.1.20

Letot, C., et al. (2015). A degradation model for maintenance improvement in respect of cost and availability. *Journal of Quality in Maintenance Engineering, 21*(1), 55–69. https://doi.org/10.1108/JQME-01-2013-0001

Liu, B., et al. (2017). Condition-based maintenance for systems with aging and cumulative damage based on proportional hazards model. *Reliability Engineering and System Safety*. Elsevier Ltd, *168*(April), 200–209. https://doi.org/10.1016/j.ress.2017.04.010

Lyubchenko, A. *et al.* (2018). An approach for optimal maintenance planning of radio communication devices considering reliability and operational costs. In *Moscow workshop on electronic and networking technologies, MWENT 2018 – Proceedings*, pp. 1–5. https://doi.org/10.1109/MWENT.2018.8337301

Mizutani, S., Zhao, X., & N. T. (2021). WIB (which-is-better) problems in maintenance reliability policies. In K. B. Misra (Ed.), *Handbook of advanced Performability engineering* (pp. 523–547). Springer. https://doi.org/10.1007/978-3-030-55732-4_23

Nasrfard, F., Mohammadi, M., & Rastegar, M. (2022). Probabilistic optimization of preventive maintenance inspection rates by considering correlations among maintenance costs, duration, and states transition probabilities, *Computers and Industrial Engineering*. Elsevier Ltd, 173(November 2021), p. 108619. https://doi.org/10.1016/j.cie.2022.108619

Sánchez Herguedas, A., Crespo Márquez, A., & Rodrigo Muñoz, F. (2022). Optimizing preventive maintenance over a finite planning horizon in a semi-Markov framework. *IMA Journal of Management Mathematics, 33*(1), 75–99. https://doi.org/10.1093/imaman/dpaa026

Sánchez-Herguedas, A., Mena-Nieto, A., & Rodrigo-Muñoz, F. (2021). A new analytical method to optimise the preventive maintenance interval by using a semi-Markov process and z-transform with an application to marine diesel engines. *Reliability Engineering and System Safety, 207*(March 2021), 1–15. https://doi.org/10.1016/j.ress.2020.107394

Sánchez-Herguedas, A., Crespo-Márquez, A., & Rodrigo-Muñoz, F. (2022a). Optimising the preventive maintenance interval using a semi-Markov process, z-transform, and finite planning horizon. In González-Prida, V., Márquez, C. A. P., and Márquez, A. C. (eds) *Cases on optimizing the asset management process* (2022nd ed., pp. 137–161). IGI Global. https://doi.org/10.4018/978-1-7998-7943-5.ch006

Sánchez-Herguedas, A., Mena-Nieto, Á., & Rodrigo-Muñoz, F. (2022b). A method for obtaining the preventive maintenance interval in the absence of failure time data. *Eksploatacja i Niezawodnosc – Maintenance and Reliability, 24*(3), 564–573. https://doi.org/10.17531/ein.2022.3.17

Sánchez-Herguedas, A., et al. (2022c). 'Optimisation of maintenance policies based on right-censored failure data using a semi-Markovian approach. *Sensors, 22*(4), 1–18. https://doi.org/10.3390/s22041432

Stogiannis, D., & Caroni, C. (2013). Issues in fitting inverse gaussian first hitting time regression models for lifetime data. *Communications in Statistics: Simulation and Computation, 42*(9), 1948–1960. https://doi.org/10.1080/03610918.2012.687061

Wang, X. (2010). *An Inverse Gaussian Process Model for Degradation Data, 52*(2), 188–197. https://doi.org/10.1198/TECH.2009.08197

Wang, J., & Miao, Y. (2021). Optimal preventive maintenance policy of the balanced system under the semi-Markov model. *Reliability Engineering and System Safety*. Elsevier Ltd, *213*(107690). https://doi.org/10.1016/j.ress.2021.107690

Wang, J., Makis, V. and Zhao, X. (2019) Optimal condition-based and age-based opportunistic maintenance policy for a two-unit series system, *Computers and Industrial Engineering*. Elsevier, 134(October 2018), pp. 1–10. https://doi.org/10.1016/j.cie.2019.05.020

Wu, B., Cui, L., & Fang, C. (2019). Reliability analysis of semi-Markov systems with restriction on transition times. *Reliability Engineering and System Safety, 190*(May), 106516. https://doi.org/10.1016/j.ress.2019.106516

Wu, B., Maya, B. I. G., & Limnios, N. (2021). Using semi-Markov chains to solve semi-Markov processes. *Methodology and computing in applied probability, 23*(4), 1419–1431. https://doi.org/10.1007/s11009-020-09820-y

Zhang, X., & Gao, H. (2012). Road maintenance optimization through a discrete-time semi-Markov decision process. *Reliability Engineering and System Safety, 103*, 110–119. https://doi.org/10.1016/j.ress.2012.03.011

Zhu, X., et al. (2021). Maintenance optimization based on three-stage failure process under performance-based contracting. *Mathematical Problems in Engineering, 2021*, 1–14. https://doi.org/10.1155/2021/6323844

Antonio Sánchez-Herguedas holds a Ph.D. in Industrial Engineering from the University of Seville (Spain), specializing in "Asset Management". He has 23 years of professional experience, fourteen of them in the Rolls Royce group. He is currently Associate Professsor in the Department of Industrial Organization and Business Management at the School of Engineering of the University of Seville. He has published numerous research papers in prestigious national and international journals and conferences and is currently a member of the research group "TEP134: Industrial Organization".

Antonio Jesús Guillén López obtained his PhD in Industrial Engineering from the University of Seville, currently holds the position of Professor in the Department of Organization of Companies at the University Complutense of Madrid. He is member of the SIM (Sistemas Inteligentes de Mantenimiento) research group at the University of Seville, contributing significantly to the field of intelligent maintenance systems. His career has been dedicated to impactful research, particularly in the areas of asset management and maintenance. His leadership extends to serving as the Secretary General of ESReDA since 2022, overseeing European Safety, Reliability, and Data Association initiatives. Since 2012, he has collaborated as a consultant with INGEMAN (Asociación para el Desarrollo de la Ingeniería de Mantenimiento) on projects with notable companies, including ADIF, ENAGAS, TALGO, AZVI, VIESGO, and ABENGOA.

Francisco Rodrigo-Muñoz holds a Ph.D. in Physical Sciences from the University of Seville (Spain), subsequently specializing in Statistics. He has 41 years of professional experience, as senior lecturer in the Department of Applied Mathematics at the School of Engineering of the University of Seville. He has published various research papers in prestigious national and international journals as International Journal of Bifurcation and Chaos, IMA Journal of Management Mathematics and Reliability Engineering and System Safety among others.

Part IV
Industry-Specific Asset Management and Other Considerations

Audit Models for Asset Management, Maintenance and Reliability Processes: A Case Study Applied to the Desalination Plant

Pablo Duque, Carlos Parra, Félix Pizarro, Andrés Aránguiz, and Emanuel Vega

Abstract Currently, the timely identification of improvements, shortcomings, and potential failures applied to maintenance has taken relevant attention from the scientific community in recent years. In order to carry out appropriate diagnosis, the employment of methods to properly measure the reliability of industrial processes has been a trend. In this work, AMORMS and AMS-ISO 55001 are applied to a seawater desalination plant aiming for carrying out a fitted measurement, generating suited improvement plans. In this context, AMORMS is a model based on 8 phases, which focuses on assets management. On the other hand, AMS-ISO 55001 focuses on the asset management norm ISO55001. The results yielded include the design and generation of actions to tackle the 20% more deficient categories needed to achieve a competitive industrial performance.

1 Introduction

Copper production in Chile has fallen between 17 and 24% due to the current drought, which end up forcing the copper companies to diversify their water sources (América económica, 2022). The Chilean Copper Commission estimates that during 2020, 73% of the water from mining processes was recirculated and that 70% of this water came from continental sources. In addition, it has been proposed, as a National Mining Policy, to reduce in the employment of continental water related to

P. Duque (✉) · C. Parra · F. Pizarro · A. Aránguiz
Technical University Federico Santa María, Department of Mechanics, Viña del Mar, Chile
e-mail: pablo.duque@usm.cl; carlos.parram@usm.cl; felix.pizarro@usm.cl; andres.aranguiz@usm.cl

E. Vega
Catholic University of Valparaíso, School of Computer Engineering, Valparaíso, Chile
e-mail: emanuel.vega@pucv.cl

A. Crespo Márquez et al. (eds.), *Advances in Asset Management: Strategies, Technologies, and Industry Applications*, Engineering Asset Management Review 3, https://doi.org/10.1007/978-3-031-52391-5_8

industrial mining process, which should not exceed 10% of the water used in 2025 (Revista Minería Chilena, 2022), thus, concluding in an increment in numbers of desalination plants. The development of such a plant has been a key issue in the improvements related to the copper industry. Greater availability is thus necessary, resulting from a combination of maintainability and reliability. The requirements for these are quality controls, safety risk controls, and environmental risk controls (Da Silva et al., 2021). In addition, failure management should be considered as a coordinated activity aimed at the prevention and resolution of failures (Schneider et al., 2019).

In this context, the necessity to identify, review, and optimize the asset management and maintenance processes regarding this type of plant has taken relevant attention in the last decade. In this work, we propose the employment of two audit processes, AMORMS (Asset Management, Operational Reliability & Maintenance Survey) and AMS-ISO 55001 (Asset Management Survey ISO 55001), in order to identify gaps in maintenance management model to a desalination plant. Regarding the results achieved, the first model classifies the plant as having "average standard processes", the second model illustrates the classification as "Processes with very good practices". Thus, we analyse and highlight the drawbacks and gaps detected. Also, action plans are designed, generated, and proposed aiming for short and medium term.

2 General Background

In the literature, maintenance audits have been proposed to carry out evaluations of different processes and areas that comprise Maintenance Management. This process, usually end up generating detailed fitted action plans aiming for improvements. The evaluation process concerns identifying the gaps, generation of recommendations in order to reduce or remove deviations (Parra Márquez et al., 2021). If we consider the Maintenance Management Model, audits are part of Phase 8: Implementation of the continuous improvement process and adoption of new technologies, as depicted in Fig. 1 (Parra Márquez et al., 2021). In this case, since it is a recently operational plant, an initial diagnostic audit is conducted within this phase.

The methodologies for auditing Maintenance Management processes are varied, but they have in common the use of questionnaires to be applied to personnel within the organization, both from operations and maintenance, in order to achieve a comprehensive evaluation. In this case study, two auditing techniques will be employed: AMORMS and AMS-ISO 55001. The main objective behind a maintenance audit is to carry out a measurement of different processes and areas that make up the Maintenance Management. Once the opportunities for improvement have been identified, action plans will be generated. The main differences between the models to be used are:

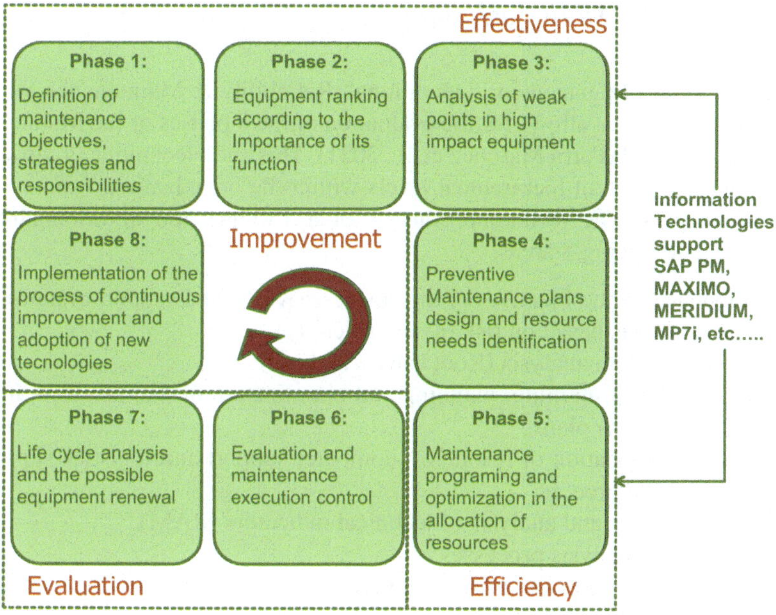

Fig. 1 Maintenance Management process model

- AMORMS is an audit which focuses on the process by evaluating the 8 phases of the Maintenance Management Model proposed by Parra and Crespo (Crespo Márquez & Parra Márquez, 2015). This audit includes some topics regarding the management model that has been proposed by the standard ISO 55001. Also, it gives special emphasis on the employment of support tools for the management process.
- AMS-ISO 55001 focuses on auditing the processes that make up the asset management system as a strategic process from a sustainable perspective throughout its lifecycle, in accordance with the definition specified in the standards set out in ISO 55000 (De Souza et al., 2022) (ISO, 2014).

AMORMS and AMS-ISO 55001 complement each other by addressing different aspects of asset management. AMS-ISO 55001 focuses on auditing the management system according to the requirements of the ISO 55001 standard, ensuring compliance with established standards and practices. On the other hand, AMORMS is responsible for auditing the use of support tools for asset management processes and their interaction with other areas of the company, such as production and supply. This synergy between both approaches is essential to gain a comprehensive view of asset management and ensure its efficiency and effectiveness within the organization.

2.1 AMORMS Audit Technique

AMORMS (Asset Management, Operational, Reliability & Maintenance Survey), as illustrated in Fig. 1, allows for the evaluation of the 8 phases of the Maintenance Management model (Parra Márquez et al., 2021). This audit technique is carried out by considering different hierarchical levels within the organization, from supervisors to managers. The technique consists of a 150-question questionnaire that addresses the following 8 areas.

- Asset management, objectives of the business (KPIS) and support organization.
- Hierarchy models based on risk (asset criticality).
- Process of problem analysis (Root cause Analysis).
- Processes of programming, planning and optimization of maintenance, inspection and operations plans.
- Processes of allocation of resources, computer support and logistics support to the maintenance process.
- Control Processes and analysis of technical indicators (RAM)
- Life cycle cost analysis processes.
- Review and continuous improvement processes.

The survey was designed to gather comprehensive information regarding these areas, enabling a thorough assessment of the organization's maintenance management practices. Thus, each area is carefully examined to identify strengths, weaknesses, and areas for improvement. The AMORMS audit technique provides a valuable framework for assessing and enhancing maintenance management processes, facilitating informed decision-making and the implementation of continuous improvement initiatives.

To each question in this survey, one of the following scores are assigned, Table 1.

Subsequently, the results can be presented in radar charts, which visually identify the points with deficiencies that should be addressed with action plans, as shown in Fig. 2.

Table 1 AMORMS audit results

Score	Description
1	Very inefficient process
2	Below average process
3	Average standard process
4	Process with very good practices
5	World-class process level

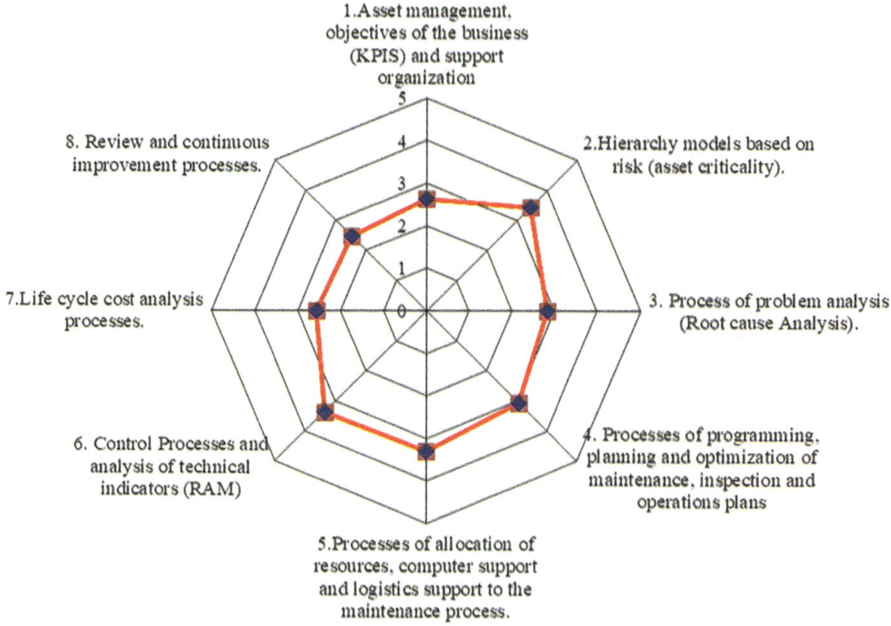

Fig. 2 Example of AMORMS audit results

2.2 *AMS-ISO 55001 Audit Technique*

AMS-ISO 55001 (Asset Management Survey-ISO 55001) allows the requirement evaluation illustrated in the ISO 55001. This audit identifies the gaps that the organization has in relation to the following requirements (Parra Márquez et al., 2021).

- Context of the organization
- Leadership
- Planning
- Support
- Operation
- Performance evaluation
- Improvement

The audit is carried out by a specially selected group of at least 10 individuals from various areas and roles within the organization to respond to the 94 audit questions of AMS-ISO 55001. The evaluation is performed employing the rating scales of existence/application, maturity level scale of Asset Management, developed by the IAM: Institute of Asset Management, see Table 2.

Table 2 AMS-ISO 55001 audit results

Score	Existence of the requirement		
0	Non-existent process	Inexperienced	0–0.5
1	Very poor process	Aware	0.6–1.5
2	Below average process	Developing	1.6–2.5
3	Average standard process	Competent	2.6–3.5
4	Process with very good practices	Optimized	3.6–4.5
5	World-class level process	Excellent	4.6–5
Score	**Application scale of requirement**		
0%	Non-existent process	Inexperienced	0–10%
20%	Very poor process	Aware	11–20%
40%	Below average process	Developing	21–40%
60%	Average standard process	Competent	41–60%
80%	Process with very good practices	Optimized	61–80%
100%	World-class level process	Excellent	81–100%

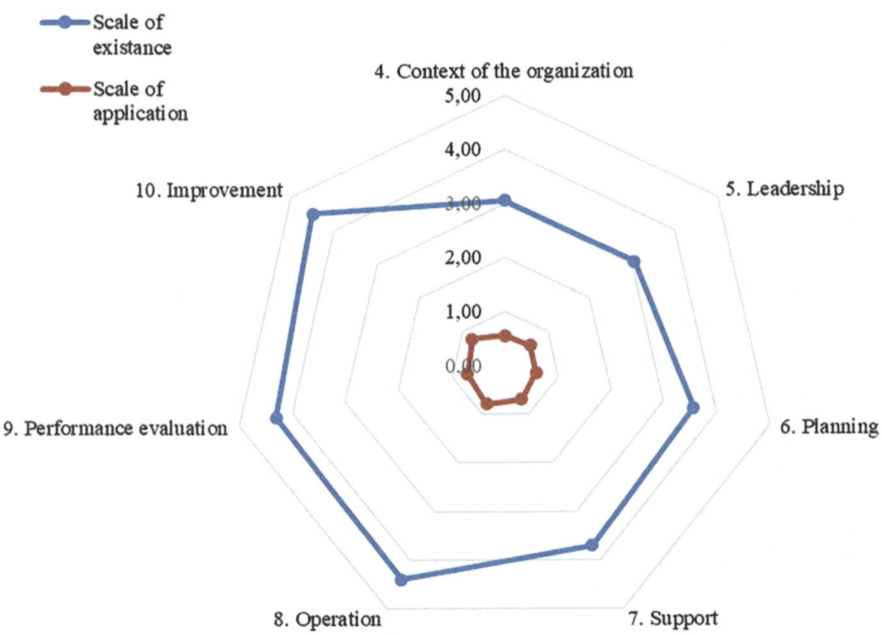

Fig. 3 Example of AMS-ISO 55001 audit results

Just like the previous technique, the obtained data is plotted in Fig. 3. Then, to determine the maturity of the organization, the maturity scale based on the requirements of the ISO 55001 standard is used, as shown in Tables 3 and 4.

Table 3 Levels of maturity of the organization regarding the requirements of ISO 55001

Score	Description	Definition	Maturity characteristics
0	Inexperienced	The organization has not recognized the need for this requirement and/or there is no evidence of commitment to implement it.	
1	Aware	The organization has identified the need for this requirement, and there is evidence of attempts to progress in implementing it.	The proposals are under development, and some requirements may be implemented. The processes are weakly controlled, reactive, and their performance is unpredictable.
2	Developing	The organization has identified the means to systematically and consistently meet the requirements and can demonstrate progress with credible plans and established resources.	The processes are planned, documented (when necessary), applied, and controlled at a local level or within functional departments, often in a reactive manner, but they may achieve the expected results repeatedly. The processes are insufficiently integrated, with limited coherence or coordination within the organization.
3	Competent	The organization can demonstrate systematic and conscious compliance with the established requirements in ISO 55001.	This level involves a formally documented asset management system that is embedded within the organization. The performance of the elements of the asset management system is continuously measured, reviewed, and improved in order to achieve the objectives of asset management.
4	Optimized	The organization can demonstrate that it is systematically and consciously optimizing its asset management practice, aligned with corporate objectives and operational context.	The characteristics of this level include the following: performance monitoring and evaluation; balancing competitive goals within an agile decision-making structure; innovation as a way of life, with continuous improvement widely demonstrated through evidence of results; reference-based improvement employed to identify additional opportunities; and a more integrated and effective management system

2.3 Background of the Audited Unit

The audited company is a copper mining and processing company located in northern Chile. Through its mining operations, it produces copper concentrate and cathodes. Its on-site infrastructure mainly includes mineral crushing and transportation systems, concentrator plants, leaching pads, solvent extraction plants, and an electrowinning plant. It also operates two pipelines that transport the concentrate to port facilities, where it is filtered and shipped to customers. Additionally, two desalination plants operate on-site, producing industrial water that supplies 95% of the plant's water requirements. This water is pumped to the mine through aqueducts.

Table 4 Levels of maturity of the organization regarding the requirements of ISO 55001 (continued)

Score	Description	Definition	Maturity characteristics
5	Excellent	The organization can demonstrate that it employs cutting-edge practices and achieves maximum value with the management of its assets, aligned with corporate objectives and operational context.	This is a dynamic and context-sensitive state, so the evidence should include demonstration of awareness of comparative assessment positions against the best similar organizations in their class, and that there are unknown improvements in asset management practices and results (value generation) that have not yet been implemented.

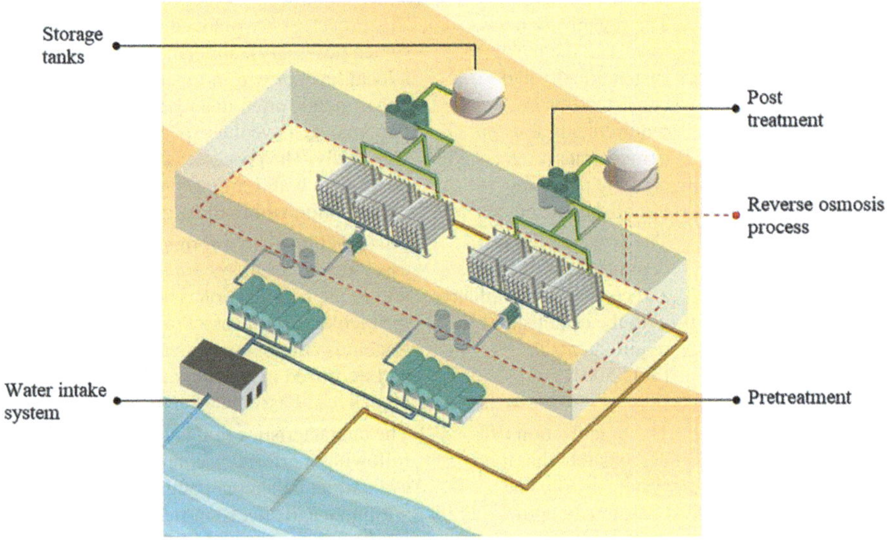

Fig. 4 Overview of a reference desalination plant

The audited area corresponds to the desalination plants, which have approximately 5000 assets according to the maintenance computer system. The desalination stations are required to achieve 95% availability and have a maintenance policy of 97% condition-based maintenance and only 3% preventive maintenance based on frequency.

The Fig. 4 shows an overview of a reference desalination plant (Servicio de Evaluación Ambiental de Chile, 2023), its main assets include:

- Water intake system: Intake tower, Outfall, Bilge
- Pretreatment: Physical-chemical pretreatment, Filtration, pH adjustment, Desanding, Microfiltration
- Reverse Osmosis process
- Storage tanks
- Post treatment

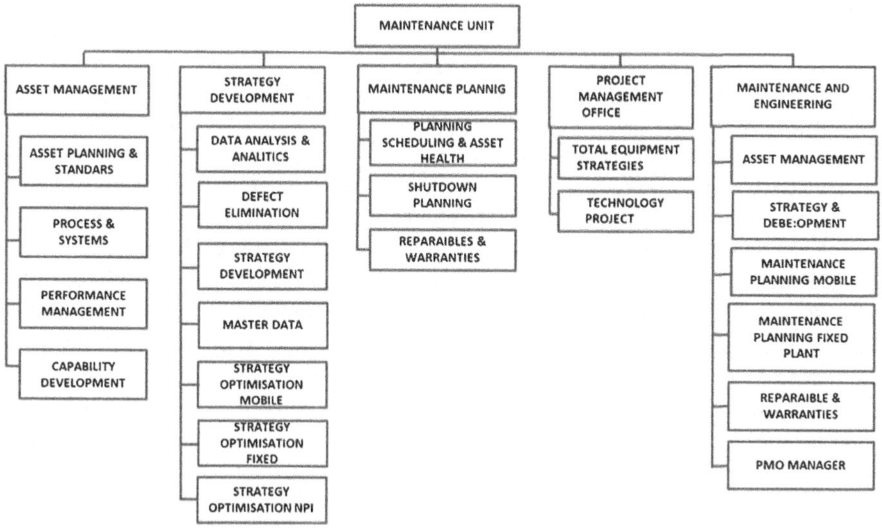

Fig. 5 Organizational diagram of the maintenance unit

The maintenance strategies and policies are supported by the following organizational structure, see Fig. 5.

Maintenance planning is carried out as follows: Guidelines are divided into 1-week guidelines for weekly inspections, 4-week guidelines for monthly maintenance, 8-week guidelines for bi-monthly maintenance, and 12-week guidelines for quarterly maintenance. There are also major maintenance guidelines for reverse osmosis racks, which occur every 36 months with Outage or Overhaul scope.

The management system used for the company's asset management is the SAP maintenance module, which handles key maintenance-related functions such as the adjustment of manpower load capacity, creation of notifications, generation of work orders, loading of master data for strategy, recording of man-hours for completed work, generation of material withdrawal reservations, and analysis and cost control. For planning and scheduling, the WMS (Work Management Scheduler) module is used, which can display activities with details such as operations, assigned workshops, duration, and associated resources.

3 Audit Results

In this section the results achieved by the audits are illustrated and discussed. In this process, planning engineers, reliability engineers, field Supervisor engineers, and maintenance superintendent have participated.

3.1 AMORMS Audit Result

In Fig. 6 and Table 5 we illustrate the results achieved by AMORMS. The description of the table is as follows, column 1 depicts each phase reviewed. The column Average, display the score achieved after carrying out the measurement of such a phase. Lastly, column Process Description, illustrates the classification given by the audits based on the score achieved. We can observe that the value illustrated in the row $\overline{X}$ (3.88), given by the 8 computed Average values, allows this plant to have a classification of "average standard processes". Also, we highlight that 37.5% of the processes reach a classification of having "very good practices", and the remaining 62.5% achieved an "average standard". In order to identify the points to improve, the Pareto Principle will be employed at two levels: Firstly, the 20% of the subcategories with the lowest score will be considered. Secondly, for each of these subcategories, the 20% of the questions with the lowest score will be considered, and action plans will be generated for the latter.

The subcategories identified within the lowest 20% achieved score are illustrated in Table 6. In this regard, the subcategories have an Average that allows them to be classified as processes with "standard average".

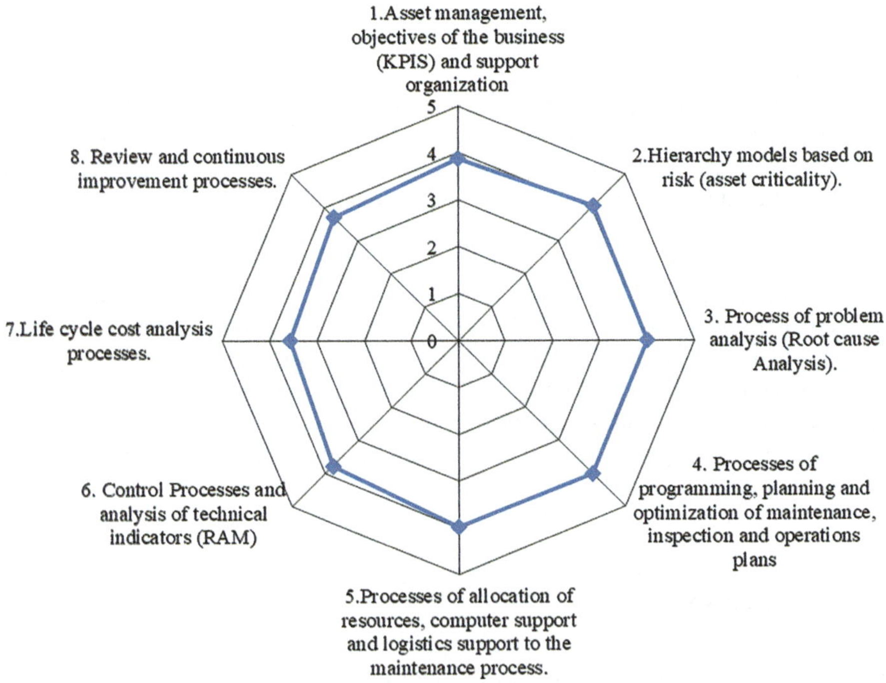

Fig. 6 Results per phase of the Maintenance Management Model, AMORMS Audit

Table 5 AMORMS audit results

Phases audited		Average	Process description
1	Definition of the maintenance objectives and KPI's	3.88	Standard average
2	Asset priority and maintenance strategy definition	4.07	Very good practices
3	Immediate intervention on high impact weak point	4.00	Very good practices
4	Design of the preventive maintenance plans	4.01	Very good practices
5	Preventive plan, Schedule & resources optimization	3.97	Standard average
6	Maintenance execution assessment and control	3.78	Standard average
7	Asset life cycle analysis	3.57	Standard average
8	Continuous Improvement and new tech	3.74	Standard average
	$\overline{X}$	3.88	Standard average

Table 6 20% of the Subcategories with lowest score according to the AMORMS Audit

Subcategories of phases of the maintenance management model		Average
1.2	Asset Management Plan	3.59
6.3	Operations control processes	3.57
6.5	Workshop management	3.63
7.1	Asset Life Cycle Cost Management	3.15
7.2	Management of information in the Asset Life Cycle	3.56
8.3	Staff development programs	3.11

In Table 7, we present the questions that yielded the lowest score, it can be seen that 3 questions depicted reached a score that categorizes them as an “Below average process”.

Tables 8 and 9 illustrates the proposed action plans designed to reduce the gaps detected, which are aligned with the company’s maintenance strategies. Thus, in order to control the progress and compliance within the action plans, indicators will be designed by the experts based on the established objectives, goals, thresholds, and definitions (AENOR, 2003). Moreover, when reviewing the answers from the surveys, it can be concluded that there are tools to support management processes, which are used systematically and consciously. However, they are not known by all the members of the Maintenance Management, so re-instruction and diffusion concerns task to be included into the planning. Regarding question 8.3.5, this was classified as a “below average process”, which gives the recommendation of carrying out an update in the training program. The proposed plan should be focused on developing the technical skills, such as knowledge, skills, and aptitudes, which are specified for each job position and stated in the given description. Also, the addition of a periodic instruction based on the deficient processes is recommended.

Table 7 20% of the questions of subcategories with lowest score according to the AMORMS Audit

#	Subcategory	FAQ	Score
1.2.2.	Comprehensive Asset Management Plan	Exists a comprehensive plan designed to implement the various processes proposed by the asset management model?	3.17
6.3.1	Operations Control Processes	Exists a procedure that details the operational processes?	3.17
6.5.3	Workshop Management	Exists a standardized contract model developed for all services requested from the workshops?	3.57
6.5.4		Exists a specific procedure in place to evaluate the delivery times, costs, and quality of execution of services provided by the workshops?	3.57
6.5.5		Exists a certified audit and benchmarking model under a local or international standard that allows the evaluation of services provided by the workshops?	3.57
7.1.5	Life Cycle Cost Analysis Processes	Is the lifecycle information of assets efficiently documented, and are the results of the lifecycle of selected equipment audited?	3.00
7.2.1	Asset Lifecycle Information Management	Does the organization's management regularly review key factors of its asset management system (including asset management policy, strategy, objectives, and plans) to ensure their effectiveness and adequacy?	2.66
8.3.3	Personnel Development Process.	Exists a specific training plan tailored to the entire worker lifecycle?	2.66
8.3.5		Does the training program include education in the areas of modern maintenance techniques, reliability, and asset management?	2.66

3.2 AMS-ISO 55001 Audit Result

According to the results illustrated in Fig. 7 and Table 10, the 7 requirements defined by the AMS-ISO 55001 audit can be classified as "Process with very good practices". This can be interpreted as the organization demonstrating being systematically and carefully optimizing their own asset management practice, which is consequence of being aligned with corporative objectives and their operating context.

As in the previous case illustrated in Sect. 3.1, the 20% of the subcategories with the lowest scores are identified, see Table 11, all of which are classified as processes "with very good practices".

In Table 12, we illustrate the results with the lowest scored as Average. In this regard, we can observe a similar situation identified by the application of AMORMS, there is a lack of knowledge related to the activities associated within the measured requirements defined by the standard ISO 55001, and how they are connected to the activities performed by them. It is proposed that the diffusion and re-training needs to be carried out at the operational, tactical, and strategic level on a regular basis.

Table 8 Proposed action plans for AMORMS audit

Subcategory – question – score	Action	Indicator
Comprehensive Asset Management Plan 1.2.2 Is there a comprehensive plan designed to implement the various processes proposed by the asset management model? **Score:** 3.17	The comprehensive plan exists, which is why it is necessary to disseminate the existing one. The dissemination will be done through training, workshops and dashboard.	**Name:** Re-instruction of the comprehensive asset management plan $= \left(\frac{total\ attendees}{total\ maintainers} \right) * 100\%$ **Goal:** 80%; **Range:** 75 to 85% **Responsible**: Superintendent Reliability
Operations control processes **6.3.1** Is there a procedure detailing the operational processes? **Score:** 3.17	The procedures exist, but they must be updated and disseminated. Dissemination will be done through training and workshops	**Name:** Procedures Update $= \left(\frac{updated\ processes}{obsolete\ processes} \right) * 100\%$ **Goal:** 80%: **Range:** 75 to 85% **Responsible:** Operations Superintendent
Workshop management **6.5.3** Is there a standard contract model developed for all the services requested from the workshops? **Score:** 3.57	The standard contract model exists, but it is necessary to train staff in the scope of existing contracts. Dissemination will be done through training and workshops	**Name:** Re instruction scope contracts $= \left(\frac{total\ attendees}{total\ maintainers} \right) * 100\%$ **Goal:** 80%: **Range:** 75 to 85% **Responsible:** Contract Administrators

Table 9 Proposed action plans for AMORMS audit (continued)

Subcategory – question – score	Action	Indicator
Workshop management **6.5.5** Is there an auditing and benchmarking model certified under a local or international standard, which allows evaluating the Services offered by the workshops? **Score:** 3.57	It exists, which is why it is necessary to re-instruct and reinforce the existence of the evaluation process for transversal services in external repair shops. Dissemination will be done through training and workshops	**Name:** Re-instruction evaluation of services $= \left(\frac{total\ attendees}{total\ maintainers}\right) * 100\%$ **Goal:** 80%; **Range:** 75-85% **Responsible:** Reliability Superintendent
Life Cycle Cost Analysis Processes **7.1.5** Is the information on the life cycle of the assets efficiently documented and are the results of the Life Cycle of the selected equipment audited? **Score:** 3.00	There is an Asset management model that incorporates the Life Cycle Cost Analysis processes, for which it is necessary to re-instruct the organization's Asset Management model, in addition to disseminating it. The dissemination will be done through training, workshops and dashboard.	**Name:** Re-instruction Asset management model $= \left(\frac{total\ attendees}{total\ maintainers}\right) * 100\%$ **Goal:** 80%; **Range:** 75-85% **Responsible:** Reliability Superintendent
Management of information in the Asset Life Cycle **7.2.1** Does the organization's management regularly review the key factors of its asset management system to ensure its effectiveness, suitability **Score:** 3.33		
Staff development process **8.3.5** Does the training program include training in the areas of modern maintenance techniques, reliability and asset management? **Score:** 2.86	Although the maintenance program includes training in the areas of modern maintenance techniques, reliability and asset management, it must be reviewed and adjusted according to the needs of each job. In the short term, the development of training in the indicated areas is proposed. Topics associated with the re-instructions requested in the previous points should also be included.	**Name:** Re-instruction maintenance techniques $= \left(\frac{total\ attendees}{total\ maintainers}\right) * 100\%$ **Goal:** 80%; **Range:** 75-85% **Responsible:** Superintendent of Personal Development and Superintendent of Execution

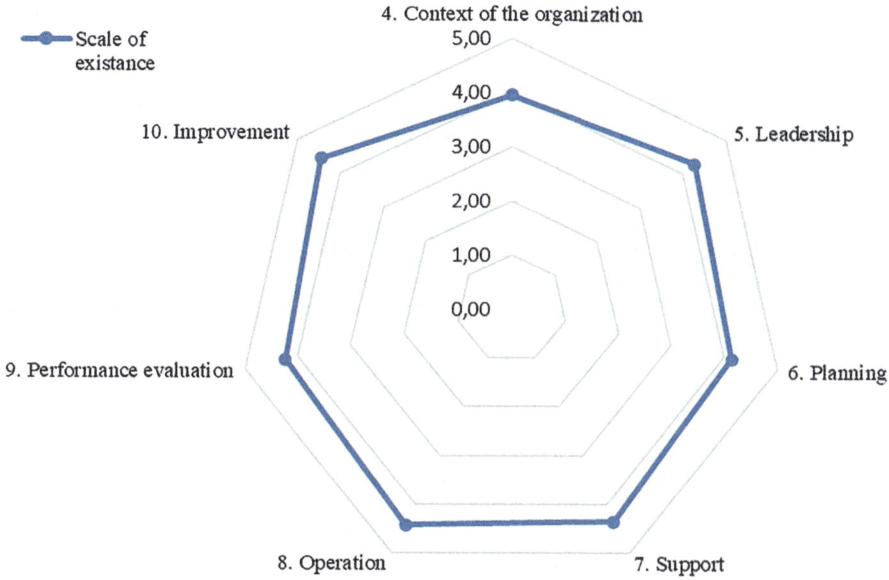

Fig. 7 Results by ISO 55001 Requirements, AMS-ISO 55001 Audit

Table 10 AMS-ISO 55001 audit results

ISO 55001 requirements		Average	Process description
4	Organizational Context	3.95	With very good practices
5	Leadership	4.29	With very good practices
6	Planning	4.14	With very good practices
7	Support	4.34	With very good practices
8	Operation	4.42	With very good practices
9	Performance evaluation	4.22	With very good practices
10	Improve	4.44	With very good practices

Table 11 20% of the Subcategories with the lowest score, AMS-ISO 55001 Audit

	ISO 55001 requirements subcategories	Average
4.1	Understand the organization and its context	3.9
4.2	Understand the needs and expectations of stakeholders	3.9
4.3	Determine the scope of the asset management system	4.0
4.4	Asset Management System	4.0
5.1	leadership and commitment	4.1

Table 12 20% of the questions of Subcategories with lowest score according to the AMS-ISO 55001 Audit

#	Subcategory	FAQ	Score
1.2.2	Asset Management Plan	Exists a comprehensive plan designed to implement the various processes proposed by the asset management model?	3.17
6.3.1	Operations Control Processes	Exists a procedure that details the operational processes?	3.17
6.5.3	Workshop Management	Exists a standardized contract model developed for all services requested from the workshops?	3.57
6.5.4		Exists a specific procedure in place to evaluate the delivery times, costs, and quality of execution of services offered by the workshops?	3.57
6.5.5		Exists a certified audit and benchmarking model under a local or international standard that allows evaluating the services offered by the workshops?	3.57
7.1.5	Life Cycle Cost Analysis Processes	Exists a lifecycle information of assets efficiently documented, and are the results of the lifecycle of selected equipment audited?	3.00
7.2.1	Asset Lifecycle Information Management	Does the organization's management regularly review key factors of its asset management system (including asset management policy, strategy, objectives, and plans) to ensure their effectiveness and adequacy?	3.33
8.3.3	Personnel Development Process	Exists a specific training plan tailored to the entire worker lifecycle?	2.86
8.3.5		Does the training program include education in the areas of modern maintenance techniques, reliability, and asset management?	2.86

The Tables 13 and 14 presents the proposed action plans designed to reduce the gaps detected.

4 Recommendations

Regarding the results yielded by both audits, it can be highlighted that the maintenance processes of the desalination unit are above average, obtaining the following overall ratings:

- According to the AMORMS audit, the processes are classified as "Average standard processes" with a score of 3.88. Through the implementation of proposed improvements, they can soon achieve a score of 4.0, categorizing them as "Processes with very good practices."
- According to the AMS ISO 55001 audit, the processes are classified as "Processes with very good practices" with a score of 4.26. They have the potential to reach the classification of "World-class processes" by achieving a score of 4.6.

Table 13 Proposed action plans for AMS-ISO-55001 audit

Subcategory – question – score	Action	Indicator
Understanding of the Organization and its context **4.1.4** Has an internal and external analysis of the key business units been carried out? **Score:** 3.43	The analyzes exist. The internal and external analysis processes of the key business units (Business Risk) must be disseminated. The dissemination will be done through training, workshops and dashboard	**Name:** Re-instruction Procedures and scope ISO 55001 $= \left(\frac{total\ attendees}{total\ maintainers} \right) * 100\%$ **Goal:** 80 %; **Range:** 75–85% **Responsible:** Reliability Superintendent
Understanding the needs and expectations of stakeholders **4.2.8** Have the requirements of the interested parties to Asset Management been identified? **Score:** 3.86	The analyzes exist. The requirements of the interested parties must be disseminated to the asset management defined by the company. The dissemination will be done through training, workshops and dashboard	
Determining the scope of the asset management system **4.3.9** Has the scope of the Asset Management System been declared, in which the main assets for the system (portfolio) were identified? **Score:** 4.00	The statements exist. The Scope of the asset management system should be disseminated. The dissemination will be done through training, workshops and dashboard.	

Table 14 Proposed action plans for AMS-ISO-55001 audit (continued)

Subcategory – question – score	Action	Indicator
Asset Management System 4.4.10 Have the commitments been established for the implementation of asset management comprehensively in all management units? **Score:** 4.00	The commitments exist. Commitments for the implementation of asset management must be disseminated comprehensively in all management units. The dissemination will be done through training, workshops and dashboard	**Name:** Re-instruction Procedures and scope ISO 55001 $= \left(\frac{total\ attendees}{total\ maintainers}\right) * 100\%$ **Goal:** 80%; **Range:** 75–85% **Responsible:** Reliability Superintendent
Asset Management System 4.4.11 Have the processes for the Business Unit Asset Management System been defined and specified? **Score:** 4.00	The definitions exist. The processes defined and specified for the asset management system must be disseminated. The dissemination will be done through training, workshops and dashboard	
Leadership and Commitment 5.1.15 Does the business unit have a strategic asset management plan (SAMP)? **Score:** 3.71	SAMP exists, it must be spread. The dissemination will be done through training, workshops and dashboard	
Leadership and Commitment 5.1.16 Has the work team in charge of the Asset Management System and reporting to senior management been designated? **Score:** 3.57	Yes it exists. The existence of the work team in charge of the Asset Management System should be publicized. The dissemination will be done through training, workshops and dashboard	

For the implementation of the proposed action plans based on the AMS-ISO 55001 audit, it is advised to conduct informative sessions evaluated by the Reliability A&I Superintendence. These sessions should address the certification requirements according to the ISO 55001 standard, highlighting the relationship between each job position and the good practices of asset management throughout their lifecycle. The development of this proposal should consider the action plans proposed based on the AMORMS Audit for the following subcategories: Comprehensive Asset Management Plan, Life Cycle Cost Analysis Processes, and Information Management in the Asset Lifecycle. This aims to save company resources and minimize disruptions to each worker's activities.

For the implementation of the proposed action plans based on the AMORMS Audit, it is suggested, among other measures already indicated, to create a formal and cross-organizational Training Program in accordance with the "Personal Development Process" (PDP) subcategory. This plan should include role-specific training and competencies for each job position, primarily in the Tactical and Operational areas, covering the entire worker's lifecycle. The PDP should include training packages that reinforce the competencies defined in the job descriptions (knowledge, skills, and aptitudes) and take into consideration the gaps identified through recurrent audit processes.

5 Conclusions and Future Work

In this work, two audits, AMORMS and AMS ISO 55001, are carried out on a desalination plant, resulting on identification of gaps in the internal process and generation of action plans to all the levels of the organization. Also, the audits yielded a classification which was based on different metrics and scores reached, which illustrated that the maintenance unit evaluated has an "Average standard process" and "Processes with very good practices", respectively.

It was determined that the main problems behind the maintenance area are directly related to the lack of knowledge of processes, plans, and management models that exist in the area. In this context, it was proposed to give a higher priority to the diffusion of knowledge through workshops, training and the use of a dashboards. Regarding the future work, we aim to keep this line of work, thus, in accordance with the results achieved by the employment of AMORMS, the objective is to perform improvements regarding the "Personnel development process". This process will be reviewed in detail and properly exploited in order to design a special training for all the technical areas in order to close the gaps and achieve world global performance all around the organization.

References

AENOR. (2003). UNE 66175:2003 Sistemas de Gestión de calidad. Guía para la implementación de sistemas de indicadores. s.l.:AENOR.

América económica. (2022). América económica. [En línea] Available at: https://www.americaeconomia.com/mineras-en-chile-cada-vez-mas-afectadas-por-la-seguia. [Último acceso: 10 noviembre 2023].

Crespo Márquez, A., & Parra Márquez, C. (2015). *Ingeniería de Mantenimeinto y Fiabilidad Aplicada en la Gestión de Activos*. Segunda ed. Asociación para el desarrollo de la Ingeniería de Mantenimiento.

Da Silva, R., de Andrade Melani, A., Carvalho Michalski, M., & Martha de Souza, G. (2021). Applying cluster analysis to support failure management policy selection in asset management: A hydropower plant case study. 31st European Safety and Reliability Conference (pp. 1838–1844).

De Souza, G., et al. (2022). *Reliability analysis and asset management* (1st ed.). The Netherlands.

ISO. (2014). ISO 55000:2014 Gestión de activos—Aspectos generales, principios y terminología. [En línea] Available at: https://www.iso.org/obp/ui#iso:std:iso:55000:ed-1:v2:es

Parra Márquez, C. et al. (2021). Técnica de Auditoría para los procesos de Mantenimiento, Fiabilidad Operacional y Gestión de Activos (AMORS & AMS-ISO 55001), s.l.: s.n.

Revista Minería Chilena. (2022). Minería Chilena. [En línea] Available at: https://www.mch.cl/2022/03/01/mineria-sustentable-conozca-las-metas-establecidas-en-la-politica-nacional-minera-2050/

Schneider, M., Lucke, D., & Adolf, T. (2019). A cyber-physical failure management system for smart factories. *Procedia CIRP, 81*, 300–305.

Servicio de Evaluación Ambiental de Chile. (2023). Servicio de Evaluación Ambiental. [En línea] Available at: https://sea.gob.cl/sites/default/files/imce/archivos/2023/04/26/Capacitaci%C3%B3n%20inicial%20Gu%C3%ADa%20para%20la%20descripci%C3%B3n%20de%20proyectos%20de%20plantas%20desalinizadoras%20en%20el%20SEIA.pdf. [Último acceso: 17 Julio 2023].

Pablo Duque The full professor and Mechanical Engineer Pablo Duque Ramírez graduated from the Federico Santa María Technical University (UTFSM), Chile. Currently, he is the head behind the careers of Industrial Maintenance Technician and Maintenance Engineering. In addition, he leads the Diploma on Welding Inspection within the UTFSM. His areas of interest include non-destructive testing, human reliability, and maintenance management. In the field of maintenance management, he has carried out applied research, focusing on maintenance audits, life cycle cost analysis, and maintenance management models which were illustrated at the World Congress on Engineering Asset Management, in 2022.

Carlos Parra is currently Professor at the Mechanical Department, UTFSM Chile. His first 18 years of experience were directly in the maintenance area in the oil, gas, manufacturing, and petrochemical sectors. He completed a master's degree in maintenance engineering (University of the Andes, Venezuela) and a Ph.D. in Industrial Engineering (University of Seville, Spain). In the last 8 years, Prof. Parra gained experience in the manufacturing, food and beverage, mining, energy and oil sectors, participating in maintenance and operations optimization projects in the main companies throughout America. Total of publications on Research-Gate: 253 (October 2023). Books published: Cases on Optimizing the Asset Management Process, 2022, IGI Global, United States and "Ingeniería de Mantenimiento y Fiabilidad aplicada en la Gestión de Activos", 2015, INGEMAN, Spain.

Félix Pizarro is a full professor in the Mechanics Department at the Viña del Mar campus of the Federico Santa María Technical University (UTFSM), Chile. He obtained his master's degree in business administration from the University of Viña del Mar in Chile (2016), with a specialization in 'Integrated Quality, Environment, and Safety Management Systems.' He held a position as a member of the Board of Overseers of the Federico Santa María Technical University. He is currently the Director of the Mechanics Department at the Viña del Mar headquarters of the Federico Santa María Technical University. He has published four research articles in reputed national and international journals.

Andrés Aránguiz is a Professor at Universidad Técnica Federico Santa María (UTFSM) in Chile, and Mechanical Civil Engineer from the University of Concepción, he holds a Master's in Asset Management from UTFSM. Currently, as the Director General of Continuing Education at the UTFSM, he has played key roles such as Head of the Industrial Maintenance Technician Program and Director of the Mechanical Department. In the industrial sector, he has been a Maintenance Engineer and Reliability Engineer, specializing in reliability modeling, developing focused maintenance plans, symptomatic maintenance, and conducting lifecycle analysis of assets, particularly in vibrational signal analysis, primarily within the Pulp and Paper industry. With significant contributions to academia, he authored two publications in Asset Management, emphasizing Maintenance Audits and action plan generation. Recognized as an 'Outstanding Teacher' in 2015 and receiving the 'Excellence in Teaching' award in 2018, he also spoke at the WCEAM 2022 conference. Currently, his efforts are devoted to enhancing Continuing Education at the Universidad Técnica Federico Santa María.

Emanuel Vega is a Full Professor of the Computer Science Department at the Pontifical Catholic University of Valparaíso (PUCV), Chile. He received both MSc in Informatic Engineering (2016) and MSc in computer science (2019) at the PUCV. Also, received his PhD degree in Computer Science from the PUCV under the supervision of Ricardo Soto (Chile) and El-Ghazali Talbi (France), in 2023. His areas of research interest include mainly Metaheuristics, Global Optimization, Machine learning, and Expert Systems. In this context, He has published about 13 scientific papers in different international conferences and journals, some of them top ranked in Computer Science, Operational Research, and Artificial Intelligence. Most of these papers are based on the resolution of real-world and academic optimization problems related to industry, manufacturing, rostering and seaports.

Audit Model for Asset Management, Maintenance and Reliability Processes: A Case Study Applied to Pulp Mill Sector

Andrés Aránguiz, Félix Pizarro, Carlos Parra, Pablo Duque, and Emanuel Vega

Abstract Currently, the optimization process in the maintenance management has been treated as a critical issue by the industry. The proposed work focuses on maintenance model diagnosis, the process aims to detect positive practices and highly possible future improvements in the models. In order to carry out the diagnostic, a systematic process is performed over the maintenance model employed through the usage of AMORMS (Asset Management, Operational Reliability & Maintenance Survey). The study case presented in this work was carried out over a pulp mill from Chile, which has an annual production over 1 million tons. Regarding the overall analysis output, several issues were illustrated in order to reach a world level performance. Thus, the employment of such instruments aims to detect key issues in urgent need to be fixed, helping in successfully designing a fitted model to be competitive and reach higher productivity.

1 Introduction

Currently, within the main companies in the forestry sector in Chile, pulp mills play a fundamental role. In this context, this industry is usually tackled by the necessity of high standards on their process, exponential demands in production, current international pulp market context, and so on. Thus, there is a high possibility to be faced against several issues and setbacks consequence of the high standards needed to properly perform.

A. Aránguiz (✉) · F. Pizarro · C. Parra · P. Duque
Technical University Federico Santa María, Department of Mechanics, Viña del Mar, Chile
e-mail: andres.aranguiz@usm.cl; felix.pizarro@usm.cl; carlos.parram@usm.cl; pablo.duque@usm.cl

E. Vega
Catholic University of Valparaíso, School of Computer Engineering, Valparaíso, Chile
e-mail: emanuel.vega@pucv.cl

A. Crespo Márquez et al. (eds.), *Advances in Asset Management: Strategies, Technologies, and Industry Applications*, Engineering Asset Management Review 3, https://doi.org/10.1007/978-3-031-52391-5_9

In this work, a leading company in the forestry sector which has a Kraft pulp plant, located in the south-central zone of Chile is analysed. This plant was designed to reach a production of 1 million tons per year, however, this threshold has never been achieved by the production plant, which was born in 2006. Also, after carrying out a measurement based in the Overall Equipment Effectiveness (OEE) in 2019, results displayed an 88.3% which differs from the 93.5% value objective. The observed difference has been interpreted as consequences of multiple issues, such as catastrophic equipment failures, low production efficiency, and a quality rate far from 100% regarding finished products. Thus, in order to design an overall optimization, the maintenance area was complemented by the incorporation of a reliability unit. The main objective was the implementation of a maintenance model, which aims at the operational risk management and strategic decision-making in the management of its assets. In this work, in order to determine the current status of the implementation and effectiveness of the maintenance management model, a systematic audit process is carried out.

In the literature, audits have successfully been applied to measure different features, components, and tasks related to the asset management field (Roda & Macchi, 2018; Parra Márquez et al., 2020a, b). They have also improved at all organizational levels the understanding of the profit behind the usage of these tools which can achieve a great impact on the business (Lima et al., 2020). On the other hand, it is well-known that the consequence of poor maintenance can end up in process failures, life risks of assets, and so on. In this regard, authors proposed a method based on the ISO 55001 and Balanced Scorecard (BSC) was employed to define maintenance performance indicators (MPIs) for asset management. In this case, the audit identified the performance evaluation requirement for the system. In this context, we employ the AMORMS (Asset Management, Operational Reliability & Survey) (Parra Márquez & Crespo Márquez, 2020), which is a tool that evaluates the 8 phases from the management model. Also, a traditional employment of such a tool may display some bias in the results or deficient visualization of key problems as generalization can be carried out in the analysis. Nevertheless, in this work, the 8 phases and subprocess are evaluated in details in order to thoroughly find potential gaps in the results obtained.

2 General Background

2.1 Strategic Choice of AMORMS for Evaluating the Maintenance Management Model

The selection of the comprehensive diagnostic tool AMORMS is grounded in its ability to assess the ongoing implementation of the 8 phases of the Maintenance Management Model at the pulp mill plant. Its alignment with specific processes in maintenance, operational reliability, and asset management within the pulp industry positions it as the logical choice to assess and enhance these aspects at the plant.

Despite the presence of various auditing methodologies such as ISO55001, IAM, IIMM, and comparative studies between these tools (Duque Ramírez et al., 2023), AMORMS stands out for its comprehensive approach to asset and maintenance management. The tool provides an in-depth evaluation of process effectiveness, particularly notable for its capability to adapt specifically to the idiosyncrasies of the pulp mill plant. This customization feature maximizes its utility, enabling a more precise and relevant assessment of the plant's specific processes. The strategic choice of AMORMS translates into a thorough evaluation aligned with the unique needs of the pulp mill plant, significantly contributing to the successful implementation of the Maintenance Management Model (Senra et al., 2017).

2.2 *Description of the Tool Used for the Audit*

The processes that control the performance concerning the management systems require the evaluation of multiple sub-processes and related factors. In this context, to the best of our knowledge there is not a defined standard application, which is consequence by the difference in the operational context between a pulp mill and other plants. This difference can be explained by differences in the demographic environmental and social conditions that greatly affect the operational context. Moreover, the employment of tools that carry out evaluation of a maintenance management model, such as an audit, should be considered only a helping tool that identify weak points in an organization. However, the process concerning the selection and application of such a tool can be defined as a high complexity task. In the literature, we can find at least 5 types of audits (Parra Márquez & Crespo Márquez, 2015). In this regard, a tool directly related to this kind of evaluation can be the AMORMS, which is a well-known audit tool that was proposed in (Crespo Márquez, 2007), focuses in the evaluation of the management model eight phases, Fig. 1. This audit includes the evaluation of 150 characteristics corresponding to the management model. The measurement behind the application of this audit can be described as follows. Firstly, the scale evaluation goes from a minimum score of 0 to a maximum of 5. On the other hand, a classification will be given based on the score achieves on each phase: (1) Very poor process, when the score achieved is between 0 and 0,99. (2) Below average process, when the score achieved is between 1 and 1,99. (3) Average standard process, when the score achieved is between 2 and 2,99. (4) Process with very good practices, when the score achieved is between 3 and 3,99. (5) Process at world class level, when the score achieved is between 4 and 5.

To properly implement the AMORMS Audit, it is essential to select collaborators with a systemic vision of the different processes involved, such as operation, maintenance, logistics, and human resources. These collaborators should receive appropriate training in the methodology, including knowledge of a glossary, relevant concepts, and methodologies. Integrating this training into the plant's organizational culture is crucial to ensure that the gathered information holds significance and allows for a comprehensive evaluation of the organization's maturity level.

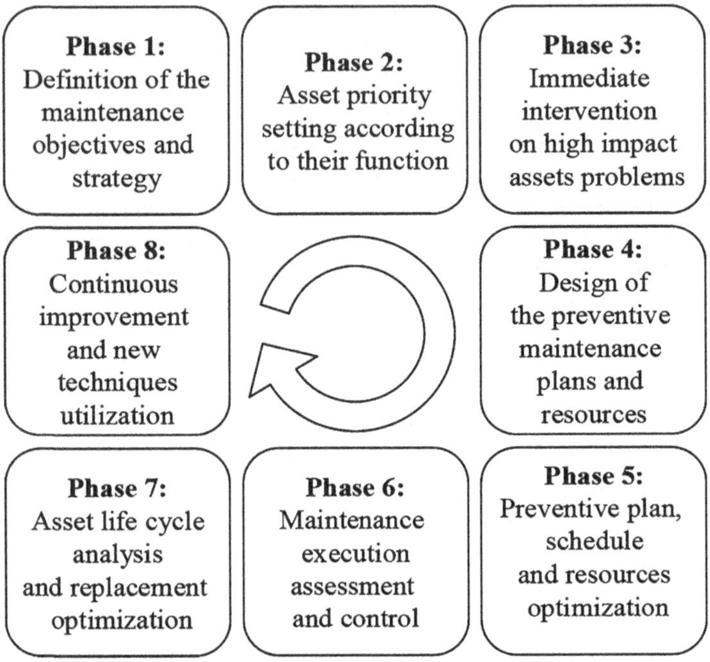

Fig. 1 Maintenance Management Model (Crespo Márquez, 2007)

The ultimate goal of this audit is to efficiently generate improvement plans that address and diminish the gaps identified in each audited phase. These improvement plans will be fundamental in optimizing the maintenance management system of the pulp mill plant, guaranteeing its optimal and effective functioning.

The final score of a characteristic can be obtained by calculating the average score across all audits for that specific characteristic. In mathematical terms, the final score (FS) of characteristic j can be expressed as:

$$C_j = \frac{\sum_{i=1}^{n_a} C_{ji}}{n_a} \tag{1}$$

Where,

C_j= Final Score of characteristic j.
n_a= represents the total number of audits conducted
j= Denotes the designation of the audited characteristic ($1 \leq j \leq 150$)
i= Represents the specific audit performed ($1 \leq i \leq n_a$)
C_{ji}=Score of charasterictic j, de la auditoría i

Similarly, each phase is evaluated by averaging each characteristic from the audit related to that particular phase. This process yields the organization's maturity level concerning the maintenance management model.

2.3 Background of the Audited Unit

Given the implementation stage of the maintenance model, the team of engineers in charge of the Plant Reliability Superintendence will be used as a sample for a preliminary evaluation of results. This group is made up of mechanical, electrical and electronic engineers, which have at least 10 years of experience in the organization. Also, engineers with less than 2 years in the organization from other industries, such as metallic mining, non-metallic mining and steel industry.

3 General Results of the Audit Process

The results obtained in the applied audit are illustrated in Table 1,which displays that the maintenance management process used in the organization is below the standard, obtaining a score of 2,7 that corresponds to an "average standard process".

The observed results can be described as follows, only 38% of the audited phases achieved a classification valued as having "very good practices", which correspond to the phases 1, 2, and 3, where the highest score was reached by phase 1. On the other hand, the remaining 62% were classified as having an "average standard process", where the lowest phase measured was phase 7. In this regard, the weakest scored group of phases will be the focus in the analysis presented in Sect. 4, the main objective is to highlight the gaps found with this tool evaluating the management process implemented in the pulp mill.

Table 1 Individual evaluation of each phase through AMORMS audit

Audited phases	Scores
Phase 1: Definition of the maintenance objectives and KPI's	3,28
Phase 2: Asset priority and maintenance strategy definition	3,00
Phase 3: Immediate intervention on high impact weak point	3,07
Phase 4: Design of the preventive maintenance plans and resources	2,79
Phase 5: Preventive plan, schedule and resources optimization	2,41
Phase 6: Maintenance execution assessment and control	2,48
Phase 7: Asset life cycle analysis and replacement optimization	2,13
Phase 8: Continuous improvement and new tech	2,48

4 Specific Results

4.1 Analysis Behind the Result Achieved on Phase 1: Definition of the Maintenance Objectives and Strategy

The sub-process evaluated in this phase and the corresponding scores achieved are illustrated in Table 2.

Considering the organizational culture of this organization, the asset management model should be implemented in this plant, as it generates the highest benefit for the holding company. The organization should clearly explain, in terms of resources and timeframes, what it plans to do with its assets. In this regard, regulations help standardize the systematization of processes, although it is not a guarantee of success. However, unlike other regulations, ISO55000, in particular, is linked to the generation of EBITDA value. Therefore, the maintenance departments must understand that asset management is the coordinated activity to generate value through assets, striking a balance between cost, performance, and risk.

Regarding the audited plant, the following observations have been noted:

- The organization has not adhered to good maintenance management practices (ISO55000, PAS55, or similar).
- The Maintenance organization lacks various elements of the maintenance management system, such as a comprehensive, well-structured, documented, communicated, and monitored strategic plan and improvement plan (covering processes, people, equipment, technology, data, information, etc.) with an action plan (responsible party, deadline, action to be taken).
- There is a general Strategic Maintenance Management Plan, but the various interviewed individuals do not thoroughly understand it.
- No Balanced Scorecard was observed that represents the connection between business KPIs and KPIs in different dimensions.

Table 2 Phase – definition of the maintenance objectives and strategy

Phase 1: definition of the maintenance objectives and strategy	scores
Managerial Vision & Leadership	3,05
Comprehensive Asset Management Plan	2,32
Comprehensive Maintenance Policies (Managerial)	3,25
Organizational Structure	3,40
Financial Control (Key Business KPIs)	3,40

- There is no traceability of previous efforts to document the maintenance strategy and processes (previous maintenance leaders or external consultants). Significant prior efforts were not implemented to define and document the maintenance policy, strategy, objectives, plans, and processes to improve the current maintenance performance.

4.2 Analysis Behind the Result Achieved on Phase 2: Asset Priority Setting According to their Function

The sub-process evaluated in this phase and the corresponding scores achieved are illustrated in Table 3.

Regarding the results of this phase, it can state that the cellulose plant has asset prioritization, which was carried out before its commissioning. However, only consequence-based prioritization was used, neglecting the frequency of failures for each asset in the plant. This has led to different initiatives. For example, in some areas, the logarithmic scatter diagram (Jack Knife) is used for prioritizing "poor performers in maintenance." In contrast, in others, only the expert judgment of former maintenance managers is used. The plant has a reasonably functional management model regarding the Health, Safety, and Environmental processes. However, this model must be constantly updated, and both management models must be integrated into a risk management model. The most significant gaps found in these phases are:

Expert judgment of operations and maintenance personnel is used for prioritization.

There is no prioritization of maintenance work considering equipment criticality, the impact of maintenance tasks, service precision, the severity of the failure, or any other parameter that aids decision-making.

To address the gap identified in this audited phase, it is crucial to establish a risk management model that incorporates a decision-making matrix for risk administration. This matrix should adopt a semi-quantitative approach, considering the failure frequency and consequence level while assigning value to the organization's high-priority parameters.

Table 3 Phase – asset priority setting according to their function

Phase 2: asset priority setting according to their function	scores
Risk Management	2,80
Equipment Prioritization	3,15
Management of Health, Safety, and Environmental Processes	3,10

4.3 Analysis Behind the Results Achieved on Phase 3: Immediate Intervention on High Impact Assets Problems

The sub-process evaluated in this phase and the corresponding scores achieved are illustrated in Table 4

The management of failures is perceived as something other than a systematic process for conducting corresponding analyses. Within the audited scope, efforts focus mainly on sporadic failures rather than chronic-acute failures. On the other hand, plant personnel are aware that these analyses add value to the production process and are trained in fault-finding tools. It should be noted that the tools vary depending on whether they are from the Operations or Maintenance area, which is reflected in the maturity level of the second subcategory. A strength present in this organization is that it has units that support failure management, such as symptomatic analysis units, which perform condition-based inspections using techniques such as ultrasound, thermography, vibration analysis, dynamic testing, oil analysis, wear analysis, etc. These inspections specifically target equipment to identify symptoms that allow potential failures to be diagnosed before they impede the asset's functionality. From the perspective of the maintenance organization's maturity profile, a "symptomatic" profile is achieved after fully reaching the "planned" profile, and it is expected that preventive maintenance policies will be complemented with predictive policies. This is because predictive policies can anticipate a failure within a reasonable time before it occurs. However, establishing and integrating a predictive policy into the organization takes work. The following considerations should be considered to reduce the gap in this phase:

- Not all reliability processes are documented.
- There is no tracking and control of root cause analysis (RCA) reports and action plans.
- The results of the analyses could be more efficiently communicated to the organization.
- Define failure modes and their symptoms before the loss of function, recognizing the measurement parameters and their operational context. Furthermore, the necessary technique(s) and measurement support or equipment should be defined as the international standards indicate (e.g. BS ISO 17359, 2018).
- Distinguish between low and high-maintainability machines, as the latter will require more complex solutions.

Table 4 Phase – immediate intervention on high impact assets problems

Phase 3: immediate intervention on high impact assets problems.	scores
Falla Management	3,05
Multidisciplinary Optimization Teams	3,25
Failure Analysis Methods	2,90

Differentiate the predictive policy's actions according to each asset's criticality within the production process. For example, critical machines with constant load and speed should have online monitoring systems.

4.4 Analysis Behind the Results Achieved on Phase 4: Desing of the Preventive Maintenance Plans and Resources

The sub-process evaluated in this phase and the corresponding scores achieved are illustrated in Table 5.

Maintenance planning is a process where materials, resources, and supplies are defined for the execution of a specific task, determining the "how" of the work. On the other hand, maintenance scheduling addresses the "who" is responsible for carrying out these tasks and "when" they should be executed (Senra et al., 2017). The planning and scheduling of all maintenance tasks, whether derived from basic maintenance plans or generated from inspections, downtime analysis, or failures, are prioritized based on their criticality.

It is important to note that there is an 18-month shutdown program in place, which has implemented operational reliability optimization techniques and maintenance procedures. These procedures clarify 31 topics, including task sequences, preparatory work, team competencies, and risk analysis when working on equipment. This gives the maintenance team a comprehensive overview of the intervention before it occurs, increasing adequate maintenance time and reducing downtime. For task execution, a working guideline is provided to guide maintenance personnel in the field, avoiding deviations, and a checklist for pre-startup equipment checks. However, these measures are only implemented for critical equipment that undergoes maintenance every 18 months, creating a need for more information for more routine maintenance tasks. Regarding scheduling, there is a cross-functional team for all five plants in the organization responsible for standardizing pre-maintenance activities. From the findings, the following can be described:

- There needs to be more documentation, such as procedures and technical information, accompanying the work plan, which hinders effective management. Not all relevant areas are always involved in weekly meetings.

Table 5 Phase – desing of the preventive maintenance plans and resources

Phase 4: desing of the preventive maintenance plans and resources	scores
Scheduling and Planning	2,85
Work Procedures and Instructions	2,50
Condition-Based Maintenance Plans (Predictive Techniques)	2,75
Optimization Techniques in Reliability, Maintenance, and Operations Areas	3,05

- There is no systematic projection of availability loss due to the maintenance program.
- Predictive plans have low visibility within the overall maintenance plans.

4.5 Analysis Behind the Results Achieved on Phase 5: Preventive Plan, Schedule and Resources Optimization

The sub-process evaluated in this phase and the corresponding scores achieved are illustrated in Table 6.

In the organization, the pulp mill has a dedicated engineering section which is in charge of the custody of plans, process diagrams and modification control, and employment of a specially parametrized software, SAP resource planning software (ERP-SAP). This last component incorporates key business functions related to the organizations, which could bring support to all the characteristics audited in this phase and have a high rating value. However, the following gaps were identified in this sub-category:

- The employment of ERP-SAP is not standardized for each area of the organization, which is why different units of the plant, such as the maintenance units, introduces default notices, work orders, route sheets, and maintenance plans are individually defined.
- Preventive and corrective maintenance plans do not have their technical and safety procedures properly loaded in the system.
- The management work orders from SAP are not suited and timely presented within the same shift, following the defined standard. There is no review in order to make sure that details, such as comments are included, data about the update of real machine hours, real man hours, backup photos, and execution reports has been an issue.
- The control in the documentation, for instance, drawings, process diagrams, and so on, is not standardized and usually depends on the guidelines of the engineering superintendent, who looks for the best approach to maintain custody of the information.

Table 6 Phase – preventive plan, schedule and resources optimization

Phase 5: preventive plan, schedule and resources optimization	scores
Maintenance computer support system (maintenance software)	2,50
Document control system	1,80
Management of spare parts, materials (logistics)	2,80
Warehouse and inventory management processes	2,55

- The handling of materials and spare parts are managed using expert judgment, without considering historical movements of materials arranged in the ERP-SAP. Thus, leading to discrepancy between the existing data.

4.6 Analysis Behind the Results Achieved on Phase 6: Maintenance Execution Assessment and Control

The sub-process evaluated in this phase and the corresponding scores achieved are illustrated in Table 7.

The main gaps detected after carrying out the audit on this phase, has been identified as consequence of deficiencies generated in the previously measured phase 5. In this context, multiple issues were detected, such as the absence of a standardized ERP-SAP employment, the use of performance indicators in the maintenance area which has not been successfully implemented to all plant areas, therefore, the following gaps were detected and highlighted:

- Unreliable information about historical equipment data in the ERP SAP.
- Absence of standardization in the indicators which evaluates the performance on maintenance areas.
- The decision making is based on expert judgment consequence by the lack of standardized indicators, thus, there is no control of the executed processes.
- Absence of a process capable to validate and audit the credentials of the contractors which participate in all the different processes relates to maintenance.

4.7 Analysis Behind the Results Achieved on Phase 7: Asset Life Cycle Analysis and Replacement Optimization

Due to the results achieved, this can be the selected phase with the greatest deviation. However, according to Table 8, it can be highlighted that the organization has a sub-process which can be classified as having "good practices related to special maintenance" with a score achieved of 3.15. This sub-process is related to Overhaul

Table 7 Phase – maintenance execution assessment and control

Phase 6: maintenance execution assessment and control	scores
Technical performance indicators	2,10
Maintenance plan review programs	2,30
Operations control processes	2,60
Contractor control	2,85
Workshop management	2,55

Table 8 Phase – asset life cycle analysis and replacement optimization

Phase 7: asset life cycle analysis and replacement optimization	scores
Asset Life Cycle Cost Management	1,50
Management of information in the Asset Life Cycle	1,75
Special maintenance (plant shutdowns, overhauls, etc.)	3,15

(BS ISO 17359, 2017), that is carried out on general plant shutdowns occurrence. They are usually controlled and guided by different units cross the entire organization, which support the planning, execution, and analysis of all the special tasks that are carried out. Also, these kinds of units have highly trained personnel to fulfil all related tasks with excellence.

On the other hand, deficient scores can be observed in the other two sub-process related to this phase. In this regard, the organization does not have a specialized support area which could give professional assistance to the execution areas, processing historical information, and facilitating decision-making for the organization's managers. Also, even if the unit which controls the reliability have been implemented, there exist a lack of specifically designed tools which calculates essentials KPIs, for instance, the TBF (time between failure) and TTR (time to repair). In this context, not having a key indicator such as the LCC (Life Cycle Cost), the units have real impediments on performing historical analysis over failures for Overhaul planning. However, if this issue were to be improved, the organization can potentially reach a higher degree of maturity on this sub-process.

On the other hand, although the organization has systems for risk management, there are many gaps in the employment of prioritization tools in different areas of the plant, which means that it is not possible to instantly analyse how the risk varies in the different areas assets of the productive areas.

The main objective behind the asset life cycle analysis is the proposition of methods which aims to evaluate different designs or actions. Thus, giving high importance to the performance of this kind of task, which propose an improvement in the efficiency related to the employment of human and economic resources in order to develop a balanced production system. The gaps detected in this phase are described as follows:

- The members of the organization are unfamiliar with the different techniques related to the analysing of the life cycle cost of assets.
- The organization's assets are not managed through life cycle cost analysis.
- The organization does not have specially designed tools with the capabilities to quantify the cost of the life cycle and make strategic decisions to reduce the risk on all the process related to production.
- Usually, most efforts from the organization are aimed to general plant shutdowns, Overhaul, and special maintenance. Thus, ignoring the potential advantages that properly managing the life cycle cost of each asset may have.

4.8 Analysis Behind the Results Achieved on Phase 8: Continuous Improvement and New Tech

The sub-process evaluated in this phase and the corresponding scores achieved are illustrated in Table 9.

Although the organization has intensified the efforts in the application of continuous improvement process, the focus behind this action can be lost when the implementation of special tools is carried out without the proper training plan. This action can be the source of several issues, which can be detrimental to different processes within the company, not generating positive changes, nor benefits within this organization. The main findings in this phase can be described as follows:

The programs and processes regarding the quality control, continuous improvement, and staff development have had support in the implementation of external advisors. However, the high rotation of key strategic professionals managing the organization has affected such definitions based on the decision-making conditioned to their expertise in the field.

Tasks related to maintenance, such as quality control and assurance processes were created in 2013, however, to date, a systematic review of said plans and processes has not been developed, due to the fact that these plans were generated by professionals outside the organization.

There is no talent management program within the organization.

5 Conclusions and Future Work

In this work, the application maintenance audit AMORMS was carried out. The output behind the analysis highlighted gaps in the maintenance management process implemented in the pulp mill. The results obtained illustrate that phase 1, 2 and 3 of the evaluated models obtained a classification of "Process with very good practices". However, phases 4, 5, 6, 7, and 8 were classified as an "average standard process". Regarding the diagnosis of this last group of phases, two main problems were observed to be the source of high potential risk issues. Firstly, the deficiencies observed in a process carried out in a specific phase can potentially be consequence of a mal functioning on other phases of the model. For instance, the gaps detected in phase 6, can be described as consequence of deficiencies belonging to phase 5. Also, we observed in phase 7 certain sub-process which were classified as "Process below average", which could have been hardly visible if a detailed evaluation and

Table 9 Phase – continuous improvement and new tech

Phase 8: continuous improvement and new tech	Scores
Quality control	2,70
Continuous improvement programs	2,85
Staff development programs	1,90

measurement on each sub-process was not properly carried out. Furthermore, it was observed that inadequate training and professional development plans for the personnel contribute to a lack of cohesion in task execution. The absence of proper training and professional development opportunities directly affects the quality and stability of the implementation of maintenance practices. These findings underscore the importance of addressing personnel management and training as critical aspects to close the identified gaps and strengthen the long-term success of the Maintenance Management Model at the pulp mill plant.

On the other hand, the final results obtained were successfully given to the organization governing and decision makers body within the pulp and paper industry, contributing to the state of the art and meeting the initial objectives established in this work. In order to give continuity and take further profit from the analysis generated, we are considering as future work the design of a strategy to improve the weakest scored phases. In this regard, we will propose a detailed improvement plan capable to reform the current organization to evolve to a model of maintenance management at the level of world-class companies.

References

BS ISO 17359. (2017). *Maintenance terminology*. BSI Standards Limited.

BS ISO 17359. (2018). *Condition monitoring and diagnostic of machines – General guidelines* (p. 108). Technical Committee ISO/TC.

Crespo Márquez, A. (2007). *The maintenance management framework. Models and method for complex system maintenance*. Springer Verlag.

Duque Ramírez, P. A., et al. (2023). Audit models for asset management, maintenance and reliability processes: A case study applied to the desalination plant. In *16th WCEAM proceedings*. Springer International Publishing.

Lima, E. S., McMahon, P., & Seixas Costa, A. P. C. (2020). Establishing the relationship between asset management and business performance. *International Journal of Production Economics*, 107937.

Parra Márquez, C., & Crespo Márquez, A. (2015). *Ingeniería de Mantenimiento y Fiabilidad Aplicada en la Gestión de Activos*. Segunda toim. Asociación para el Desarrollo de la Ingeniería de Mantenimiento.

Parra Márquez, C., & Crespo Márquez, A. (2020). *Audit techniques applied in the maintenance and reliability management processes.*, s.l.: INGECON.

Parra Márquez, C., González-Prida, V., Candón, E., De la Fuente, A., Martínez-Galán, P., & Crespo, A. (2020a). Integration of asset management standard ISO55000 with a maintenance management model. In A. Crespo Márquez, D. Komljenovic, & J. Amadi-Echendu (Eds.), *14th WCEAM Proceedings* (WCEAM 2019. Lecture Notes in Mechanical Engineering). Springer.

Parra Márquez, C., Crespo Márquez, A., González-Prida, V., Sola Rosique, A., Kristjanpoller Rodríguez, F., & Viveros Gunckel, P. (2020b). Audit models for asset management, maintenance and reliability processes. Case study: Electricity transmission sector. *Dyna Management, 8*(1), 14.

Roda, I., & Macchi, M. (2018). A framework to embed asset management in production companies. *Proceedings of the Institution of Mechanical Engineers Part O: Journal of Risk and Reliability, 232*(4, SI), 368–378.

Senra, P., Lopes, I., & Oliveira, J. A. (2017). Supporting maintenance scheduling: A case study. *Procedia Manufacturing, 11*, 2123–2130. Osa/vuosikerta.

Andrés Aránguiz is a Professor at Universidad Técnica Federico Santa María (UTFSM) in Chile, and Mechanical Civil Engineer from the University of Concepción, he holds a Master's in Asset Management from UTFSM. Currently, as the General Director of Continuing Education at the UTFSM, he has played key roles such as Head of the Industrial Maintenance Technician Program and Director of the Mechanical Department. In the industrial sector, he has been a Maintenance Engineer and Reliability Engineer, specializing in reliability modeling, developing focused maintenance plans, symptomatic maintenance, and conducting lifecycle analysis of assets, particularly in vibrational signal analysis, primarily within the Pulp and Paper industry. With significant contributions to academia, he authored two publications in Asset Management, emphasizing Maintenance Audits and action plan generation. Recognized as an 'Outstanding Teacher' in 2015 and receiving the 'Excellence in Teaching' award in 2018, he also spoke at the WCEAM 2022 conference. Currently, his efforts are devoted to enhancing Continuing Education at the Universidad Técnica Federico Santa María.

Félix Pizarro is a professor in the Mechanics Department at the Viña del Mar campus of the Federico Santa María Technical University (UTFSM), Chile. He obtained his master's degree in business administration from the University of Viña del Mar in Chile (2016), with a specialization in 'Integrated Quality, Environment, and Safety Management Systems.' He held a position as a member of the Board of Overseers of the Federico Santa María Technical University. He is currently the Director of the Mechanics Department at the Viña del Mar headquarters of the Federico Santa María Technical University. He has published four research articles in reputed national and international journals.

Carlos Parra is currently Professor at the Mechanical Department, UTFSM Chile. His first 18 years of experience were directly in the maintenance area in the oil, gas, manufacturing, and petrochemical sectors. He completed a master's degree in maintenance engineering (University of the Andes, Venezuela) and a Ph.D. in Industrial Engineering (University of Seville, Spain). In the last 8 years, Prof. Parra gained experience in the manufacturing, food and beverage, mining, energy and oil sectors, participating in maintenance and operations optimization projects in the main companies throughout America. Total of publications on Research-Gate: 253 (October 2023). Books published: Cases on Optimizing the Asset Management Process, 2022, IGI Global, United States and "Ingeniería de Mantenimiento y Fiabilidad aplicada en la Gestión de Activos", 2015, INGEMAN, Spain.

Pablo Duque The professor and Mechanical Engineer Pablo Duque Ramírez graduated from the Federico Santa María Technical University (UTFSM), Chile. Currently, he is the head behind the careers of Industrial Maintenance Technician and Maintenance Engineering. In addition, he leads the Diploma on Welding Inspection within the UTFSM. His areas of interest include non-destructive testing, human reliability, and maintenance management. In the field of maintenance management, he has carried out applied research, focusing on maintenance audits, life cycle cost analysis, and maintenance management models which were illustrated at the World Congress on Engineering Asset Management, in 2022.

Emanuel Vega is a Professor of the Computer Science Department at the Pontifical Catholic University of Valparaíso (PUCV), Chile. He received both MSc in Informatic Engineering (2016) and MSc in computer science (2019) at the PUCV. Also, received his PhD degree in Computer Science from the PUCV under the supervision of Ricardo Soto (Chile) and El-Ghazali Talbi (France), in 2023. His areas of research interest include mainly Metaheuristics, Global Optimization, Machine learning, and Expert Systems. In this context, He has published about 13 scientific papers in different international conferences and journals, some of them top ranked in Computer Science, Operational Research, and Artificial Intelligence. Most of these papers are based on the resolution of real-world and academic optimization problems related to industry, manufacturing, rostering and seaports.

The Role of Eco-Driving and Wearable Sensors in Industry 4.0

Turuna S. Seecharan

Abstract This study investigates the relationship between drivers' electrodermal activity (EDA) and their eco-driving behaviours through real-time monitoring. Electrodermal activity, a physiological marker of sympathetic nervous system arousal, reflects emotional and cognitive states, providing a valuable window into drivers' internal experiences. EDA and driving data were collected for 48 trips from 10 different drivers. Cluster analysis and the Pearson correlation coefficient was used to uncover potential patterns between driver EDA and their driving behaviour as measured using a driving score. The results follow the Yerkes-Dodson Law. Driving performance increase with EDA arousal, but only to a point. The investigation has implications for enhancing road safety, as it contributes to our understanding of how drivers' emotional states influence their on-road performance. Additionally, it holds promise for developing innovative in-car systems that can adapt to drivers' changing emotional states, promoting safer and more comfortable driving experiences. Ultimately, this study bridges the gap between psychophysiology and transportation, shedding light on the often-overlooked emotional aspects of driving behaviour.

Keywords Eco-driving · Industry 4.0 · Wearable sensors · Electrodermal activity

T. S. Seecharan (✉)
Department of Mechanical & Industrial Engineering, University of Minnesota Duluth, Duluth, MN, USA
e-mail: tseechar@d.umn.edu

A. Crespo Márquez et al. (eds.), *Advances in Asset Management: Strategies, Technologies, and Industry Applications*, Engineering Asset Management Review 3, https://doi.org/10.1007/978-3-031-52391-5_10

Table 1 Paper nomenclature

Variable	Definition
AccX	Acceleration forwards or braking
AccY	Acceleration side to side
AccZ	Acceleration up and down
$AccXPos_i$	The ith positive acceleration recorded from the Geotab G09 device.
$AccXNeg_i$	The ith negative acceleration (braking) recorded from the Geotab G09 device.
$AccYPos_i$	The ith right turn force recorded from the Geotab G09 device.
$AccYNeg_i$	The ith left turn force recorded from the Geotab G09 device.
Brake	The braking score.
C	A self-reported "calm" mood.
CDC	Centers for Disease Control
Driving Score	A weighted average of the speeding, acceleration, braking, right, and left cornering scores.
EDA	Electrodermal Activity – The variation of electrical characteristics of the skin due to perspiration or sweat gland activity.
E4 Sensor	The wristband, developed by Empatica, worn by drivers to record their EDA while driving. It has a sampling frequency of 4 Hz.
F	A self-reported "fatigued" mood.
G09 Device	The telematic device, developed by Geotab, used in the study to record driving data. It is plugged into the OBD II port of the drivers' personal vehicle.
GPS	Global Positioning System
H	A self-reported "happy" mood.
$L_S^{'}$	Length of the trip not spent speeding.
L	Length of a trip.
Left	The left cornering score.
Max EDA	The maximum EDA.
Mean EDA	The average EDA.
Med EDA	The median EDA.
Mood 1	The self-reported mood prior to driving.
Mood 2	The self-reported mood during driving.
NAcc	The total number of recorded acceleration events.
NBrk	The total number of recorded negative acceleration (braking) events.
NLCrn	The total number of recorded left turn events.
NRCrn	The total number of recorded right turn events.
OBD II	On-Board Diagnostics II – The second generation of on-board self-diagnostic equipment.
Right	The right cornering score.
S	A self-reported "stressed" mood.
Skew EDA	The skewness of all EDA values.
SpdFreq	The number of speeding events recorded by the telematic device.
Speed	Speed score
US	United States
Sd EDA	The standard deviation of all EDA values.
μS	microSiemens – Measures of skin conductance are expressed in units of microSiemens.

1 Introduction

A study performed by the Centers for Disease Control and Prevention (CDC) compared the United States (U.S.) to 28 other high-income countries regarding road safety and found that the U.S. experienced more motor-vehicle deaths than any other country with the highest rate of motor-vehicle deaths per 100,000 population (Yellman & Sauber-Schatz, 2022). The report further states that motor vehicle injuries are the leading cause of preventable death in the world, accounting for nearly 1.3 million deaths. Aggressive driving behaviour is one of the main reasons for crash risk (Office of Traffic Safety, 2021).

The National Highway Traffic Safety Administration (NHTSA) defines aggressive driving as "the operation of a vehicle in a manner that endangers or is likely to endanger persons or property" (Stuster, 2004). Aggressive driving is solely due to human decision-making and involves following too closely, driving at excessive speeds, weaving through traffic, and running stop lights and signs. In asset management, accidents resulting from aggressive driving can lead to substantial costs, including vehicle repairs, medical expenses, legal fees, and increased insurance premiums. Reducing aggressive driving, by supporting the driver, can help mitigate these costs and protect the financial health of the fleet.

Eco-driving, defined as an energy-efficient use of vehicles through less aggressive driving style, has garnered significant interest in the literature for its reported benefits on reducing aggressive driving habits thus, potentially, increasing road safety. Since eco-driving depends on the driver making the decision to engage in an eco-driving style, it becomes imperative to comprehend the underlying factors that influence driver's behaviour and performance.

In terms of the physiological state of the driver, fatigue is already known to impact safety since drivers' reaction times, awareness of hazards, and ability to sustain attention all worsen (Meng et al., 2015). Driver stress has emerged as a significant concern, directly impacting safety, operational efficiency, and driver well-being. In a search of the Scopus database of "the relationship between driver's emotional state and aggressive driving", only five papers were found from 2006 to 2022 with no studies that investigate the relationship between emotional arousal of the driver and aggressive driving using naturalistic driving data. The objective of this study is to investigate the relationship between drivers' emotional arousal (measured using Electrodermal Activity) and their eco-driving score.

2 Literature Review

Aggressive driving habits, such as rapid acceleration and frequent braking, consume more fuel and decrease fuel efficiency. By encouraging smoother driving behaviours, fleet managers can reduce fuel costs and environmental impact. Aggressive driving also places additional stress on vehicles, leading to increased wear and tear on components such as brakes, tires, and engines. A reduction in aggressive driving can extend the lifespan of fleet vehicles, resulting in lower maintenance and replacement costs.

Ecological driving ("Eco-Driving") is a term used to describe "a driving behavior (or a driving style) that aims at saving fuel and reducing harmful GHG emissions" (Andrieu & Pierre, 2012; Fafoutellis et al., 2020; Barkenbus, 2010). Eco-driving involves accelerating moderately, anticipating traffic flow and signals to avoid sudden starts and stops, maintaining an even driving pace, driving at or safely below the speed limit, and eliminating excessive idling (Barkenbus, 2010). The advantages of eco-driving go beyond CO_2 reductions to include reducing the cost of driving to the individual and producing tangible and well-known safety benefits (fewer accidents and traffic fatalities) (Barkenbus, 2010; Zarkadoula et al., 2007; Beusen et al., 2009). It is already established that fatigue negatively impacts driving ability (Al-Mekhlafi et al., 2020). Recent research investigates the role of emotions and personality traits in the occurrence of aggressive driving habits.

2.1 Emotion and Aggressive Driving

Three related research aspects can be identified when studying emotions in the car: (1) the effect of emotions on aggressive driving, (2) the detection of emotions using psycho-physiological sensors, and (3) in-car responses to regulate and influence driver emotions (Hassib et al., 2019). Aggressive personality types tend to engage in more aggressive driving behaviour (Beanland et al., 2014; Alavi et al., 2017).

Primary driving tasks include all necessary tasks that control the movement of the vehicle such as steering, accelerating, braking, and speeding. These primary tasks are strongly related to safe driving and can be negatively impacted by negative emotions (Hassib et al., 2019).

Affective state changes in a person are always accompanied by significant physiological responses such as blood flow, changes in heart rate, muscles, facial expressions, and voice. According to Russell's model, each affective state can be represented by two dimensions: arousal and valence (Russell, 1980). Arousal indicates the level of a person's involvement in reaction to a stimulus. Valence defines the positive or negative emotional state. The Yerkes-Dodson Law (Yerkes & Dodson, 1908) and the inverted U-shape model provide theoretical foundations for understanding the complex interplay between stress and performance. These models propose an optimal stress zone where driver performance is at its peak.

A study by Eboli et al., 2017, used a questionnaire to investigate the relationship between driving style and drivers' somatic, behavioural, and emotional conditions (Eboli et al., 2017). They found that a driver inclines toward a more cautious driving style when tired, sleepy, sick, or bored while driving. If the driver is gloomy, worried, nervous, or angry, they driver inclines towards a more aggressive driving style.

Another study by Ahmed et al., 2022, used an emotional intelligence (EI) survey and the Dula Dangerous Driving Index survey to analyse dangerous driving behaviour among 615 non-commercial US drivers (Ahmed et al., 2022). They found significant associations between dangerous driving behaviours and EI. Specifically, higher EI scores engaged in less dangerous driving behaviours, resulting in fewer crashes and fatalities.

(Britt & Garrity, 2006) asked participants to recall a recent time when they experienced three different anger-provoking events when driving. They then rated their behaviours and emotions during the event, and their attributions for why the event occurred. Hostile and blame attributions predicted aggressive behaviour and anger.

In a study by Lee and Winston, a simulation was used to induce negative emotional states in young drivers to examine the relationship between emotional states and driver reactions (Lee & Winston, 2016). Self-reported data were collected from 33 young driver participants who reported their emotional states at four time points during the protocol. These data were then matched with vehicle control behaviours based on measures derived from the simulator. The simulated traffic situations resulted in emotional fluctuations over time, with a positive correlation between the magnitude of negative emotions and the number of unsafe behaviours.

An anonymous, web-based survey of 769 college students was conducted at a large East Coast university to investigate the relationship between distress tolerance and risky and aggressive driving (Beck et al., 2014). The authors define distress tolerance as "the individual's capability to experience and endure negative emotional states". Driver participants self-reported their emotional states at four time points during the protocol. The authors found that, after controlling for age, gender, race, ethnicity, year in school, grade point average, and driving frequency, distress tolerance was significantly inversely related to reported risky driving and aggressive driving.

Asset managers must consider the delicate balance between stress-induced arousal and optimal performance, as excessively high or low stress levels can lead to suboptimal driving behaviours. High levels of stress can impair a driver's ability to focus, react quickly, and make sound decisions on the road. Stressed drivers may be more prone to accidents, endangering themselves, other road users, and the company's assets. Paschalidis et al. (Paschalidis et al., 2019) developed a car-following model that explicitly accounts for the stress level of the driver and quantifies its impact on acceleration-deceleration decisions. They found that drivers with higher levels of stress (as manifested in the physiological responses) express similar characteristics to the "aggressive" drivers used in some microsimulation tools. The ability to describe the behaviours of drivers, even before they may be consciously aware of their likely behaviours, will provide a significant advancement to the transportation infrastructure (Dehzangi & Williams, 2015).

2.2 *Wearable Sensors and Telematic Devices*

Industry 4.0 aims to design machines to assist humans in being more efficient. It creates cyber-physical systems, which represent tight interaction and coordination between computational and physical resources within a smart factory (Hermawati

& Lawson, 2019). Researchers have identified eight categories in which Industry 4.0 technologies can assist operators in human-cyber physical systems: (1) operators and powered exoskeletons, (2) operators and augmented reality, (3) operators and virtual reality, (4) operators and wearable trackers, (5) operators and social networks, (6) operators and collaborative robots, (7) operators and big data analytics, and (8) operators and intelligent personal assistants (Romero et al., 2016). This work shows an application of the fourth category to monitor driver stress while driving using wearable sensors, telematics, and data insights.

Stress is a dynamic process that reflects the brain's response to internal and external factors (Butler, 1993) and is defined as "a reaction from a calm state to an excited state for the purpose of preserving the integrity of the organism" (Healey & Picard, 2005). It is linked to impaired decision-making capabilities (Baddeley, 2000), decreased situational awareness, and degraded performance, which can impair driving ability. Stress is measured via cortisol levels (Hellhammer et al., 2009) or via self-reports such as the Perceived Stress Scale (PSS) (Cohen et al., 1983). These methods cannot be used to measure stress continuously for an extended period and sometimes require a person to go to a clinician or psychologist (Mishra et al., 2020).

Recent improvements in sensing capabilities and wearable sensors (E4 Empatica device) have enabled continuous detection and monitoring of stress in several conditions: controlled, semi-controlled, and free-living conditions (Gjoreski et al., 2016; Mishra et al., 2018). One study has shown student pilots to have high EDA values during highly demanding tasks (Vallès-Català et al., 2021), as highly demanding tasks put extra pressure on them. Another study used wearable sensors to measure electroencephalography/electromyography (EEG/EMG) and heart rate to evaluate driving performance while driving under stressful conditions (Hassib et al., 2019). In addition to achieving an accuracy of 78.9% for classifying valence and 68.7% for arousal, the researchers observed enhanced driving performance when ambient lighting was introduced to calm the drivers. This indicates that wearable sensors can be used to predict emotional arousal accurately.

The E4 Empatica wristband, which includes an electrodermal activity (EDA) sensor, will be used to collect the EDA values of the participants while driving. It is an innocuous device designed to acquire information in real time and continuously throughout daily activities.

The goal of this study is to investigate whether drivers drive worse when stressed. This is an observational study in which drivers wear an Empatica E4 wristband with a telematic device plugged into the On-Board Diagnostics (OBD II) port of their personal vehicle while driving. Their eco-driving performance is measured using a driving score (T. Seecharan, 2022). A survey was used to obtain the drivers' self-reported assessments of their moods. Descriptive statistics are used to search for patterns between (1) the drivers' self-reported moods and their driving scores and (2) the drivers' raw EDA and their driving scores.

2.3 Summary

Related work shows that emotions can impact driving performance. Eco-driving can potentially improve road safety by reducing hard acceleration, hard braking, and speeding. Researchers have investigated the use of physiological sensors to understand driver emotions. However, limited research investigates the relationship between EDA and driving performance and none investigate the relationship between EDA and eco-driving performance. In this work, wearable sensors are used to observe the relationship between drivers' EDA and their eco-driving performance.

3 Methodology

Analysing EDA and eco-driving involves a combination of data collection, processing, and interpretation. A general outline of the steps involved in this analysis is as follows:

3.1 Data Collection

Driving Behaviour Data: To analyse driving behaviour, data can be collected through various sources, such as vehicle telematics, GPS devices, accelerometers, or smartphone apps. The Geotab G09 device, plugged into the drivers' on-board diagnostics (OBD II) port, collected speed, acceleration, and braking patterns.

1. EDA Data: Electrodermal activity measures electrical conductance on the skin's surface, commonly known as skin conductance or galvanic skin response. The Empatica E4 device was used to collect EDA data from the drivers.
2. Data Synchronization - The EDA data and driving behaviour data must be synchronized correctly so that both datasets can be analysed in relation to each other. In the Geotab cloud, trip start and end dates along with trip lengths were recorded. These data were matched to the EDA timestamp.
3. Preprocessing: Trips less than 5 miles in length were removed, and any EDA data that were abnormally high were removed. For one driver, there was a trip in which their EDA was in the range of 30 μS. This was abnormally high for this driver and was removed from the analyses. Driving data, EDA data, and survey responses that matched in terms of trip date and duration were retained for analysis.
4. EDA Analysis – This work uses the raw EDA data in the analysis. Descriptive statistics for the EDA recorded for each trip for each driver were calculated including mean EDA, standard deviation of the EDA (sd EDA), median EDA

(Med EDA), maximum EDA value (Max EDA), and skewness of the EDA (skew EDA). The independent variable in this study was mean EDA.
5. Eco-Driving Behaviour Analysis – Eco-driving behaviour is quantified using a "driving score". Metrics to calculate this driving score are harsh acceleration, harsh braking, sharp turns, and excessive speed. This is the dependent variable in the study.
6. Correlation and Patterns - The correlation between mean EDA and driving score was examined. Hierarchical clustering was used to create separable clusters and observe differences in mean EDA, median EDA and driving score between clusters.
7. Interpretation – The results were interpreted, and conclusions about the connection between emotional arousal (EDA) and eco-driving behaviour were drawn. The implications of the findings for improving road safety, driver behaviour, and potential interventions were considered.

3.2 Participants

This paper presents the results from ten drivers recruited from the undergraduate student population at the University of Minnesota Duluth. Drivers must hold a valid driver's license and valid vehicle insurance to be included in the study. They were asked to record their EDA and driving data for five trips of at least five miles in length. Drivers were also asked to record their mood via a survey. Driving data were collected using the Geotab G09 telematics device plugged into the on-board diagnostics port of the drivers' personal vehicle. The drivers wore an E4 Empatica device while driving to record their EDA data. The E4 sensor and G09 device are shown in Fig. 1.

Driving data from the G09 device were downloaded from Geotab's cloud storage and analysed using R. The participants were also asked to complete a short survey after each driving session. The survey questions are shown below. This survey assists in matching EDA data with vehicle engine data.

The EDA from the Empatica E4 was measured with dry electrodes that detect changes in the electrical conductivity of the skin. It sampled at a frequency of 4 Hz,

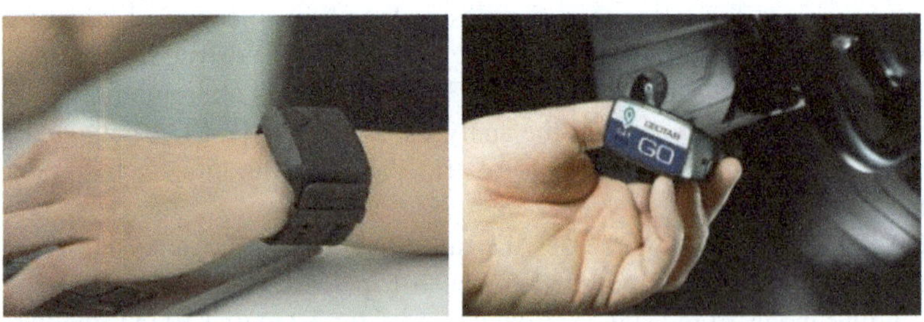

Fig. 1 E4 wristband (left) and G09 telematic device (right)

and data were measured in microSiemens (μS). The Empatica E4 is a wearable sensor worn on the wrist that is used to record physiological signals. It offers two modes of recording: (1) real-time via an app or (2) locally stored data on the device. This work used the real-time mode of recording. After finishing the real-time recording, the data were transferred to Empatica Connect via a Wi-Fi internet connection. On Empatica Connect, the E4 data can be visualized, deleted, or downloaded. Empatica offers physiological signals in raw format (e.g., EDA, blood volume pulse, temperature, and movement) but offers no tools for signal analyses.

3.3 Trip Survey

1. Participant ID?
2. Trip Date and Time?
3. Which word best describes your mood **before** your trip started?
   a) Happy
   b) Calm
   c) Stressed
   d) Fatigued
   e) Angry
4. Which word best describes your mood **during** your trip?
   a) Happy
   b) Calm
   c) Stressed
   d) Fatigued
   e) Angry
5. Please select which of the following events happened while you were driving.
   a) Sudden braking to avoid a pedestrian/cyclist/car
   b) Hostile behaviour from another driver
   c) Accident
   d) Heavy traffic
   e) None of the above (uneventful)

3.4 Driving Score

The telematic device records GPS and engine data for each driver. The engine data include acceleration forwards and braking (AccX), acceleration side to side (AccY), acceleration up and down (AccZ), GPS location, trip distance and number of times the driver "speeds" along with the distance and time spent speeding. Acceleration data are recorded at small increments in time, as shown in Table 2. For example, in the third row of Table 2, for the driver identified as "BB", at 2:34 pm in their first recorded trip, a braking event was recorded at −2.30 m/s^2.

Speeding, on the other hand, is a user-defined "rule" within the portal. A sample speeding report is shown in Table 3. For example, for the driver identified as "BB", during their 5th trip, at 2:50 pm, the driver was speeding for 0.6879 miles.

The driving score penalizes higher levels of acceleration, braking, cornering and speeding (T. S. Seecharan, 2021). To calculate the driving score, three levels were created for acceleration, braking and cornering to incorporate mid-range driving. Thresholds were chosen based on previous research on the effect of hard acceleration on vehicle fuel economy and passenger safety (Boodlal & Chiang, 2014). A speeding event for a driver depends on the posted speed limit of the road; therefore, a mid-range level for speeding was not designed. Instead, the trip length was recorded along with the length of time spent speeding.

Telematic devices collect continuous driving data and report them as discrete data at small time increments. In a trip – defined as from when the driver starts the car, drives, and then turns off the car –acceleration, braking, left cornering, right cornering and car speed are discrete values. Each discrete recording of acceleration, braking, left cornering, or right cornering is termed an "event". Each positive acceleration event is defined as $\boldsymbol{AccXPos_i}$, each negative acceleration event is $\boldsymbol{AccXNeg_i}$, each right turn event is $\boldsymbol{AccYNeg_i}$ and each left turn event is $\boldsymbol{AccYPos_i}$. In one trip, depending on the length, there are many of these events. The scoring system for

Table 2 Sample acceleration report

DriverID	TripID	time	description	value
BB	1	2:34:56 PM	AccX	0
BB	1	2:34:56 PM	AccY	0
BB	1	2:34:56 PM	AccY	−2.30
BB	1	2:35:02 PM	AccX	0
BB	1	2:35:02 PM	AccY	0
BB	1	2:35:39 PM	AccX	0
BB	1	2:35:39 PM	AccY	0
BB	1	2:35:39 PM	AccX	2.48
BB	1	2:35:42 PM	AccX	0
BB	1	2:36:05 PM	AccX	0
BB	1	2:36:05 PM	AccY	0
BB	1	2:36:06 PM	AccY	−2.83

Table 3 Speeding distance recorded for driver "BB"

DriverID	TripID	time	Distance (miles)
BB	5	14:50:53	0.688
BB	5	14:53:40	0.592
BB	5	14:54:30	1.28
BB	5	14:56:50	0.495

each metric is shown in Table 4. The thresholds were chosen from GPS tracking companies' websites (linxup, n.d.; Broughall, 2020).

The value of the event is checked against the threshold. For acceleration, braking, left cornering, and right cornering, each event is assigned a value of 0, 1, or 2 depending on its comparison to the thresholds. Using these values, the acceleration, braking, right cornering, and left cornering scores are calculated using Eqs. (1), (2), (3), and (4), respectively.

$$Accel = \left(\frac{\sum_{i=1}^{NAcc} AccXPos_i}{2 \times NAcc} \right) \times 10 \tag{1}$$

$$Brake = \left(\frac{\sum_{i=1}^{NBrk} AccXNeg_i}{2 \times NBrk} \right) \times 10 \tag{2}$$

$$Right = \left(\frac{\sum_{i=1}^{NRCrn} AccY_i}{2 \times NRCrn} \right) \times 10 \tag{3}$$

$$Left = \left(\frac{\sum_{i=1}^{NLCrn} AccY_i}{2 \times NLCrn} \right) \times 10 \tag{4}$$

In one trip, there will be a total number of acceleration events labelled ("NAcc"); a total number of braking events labelled ("NBrk"), a total number of right turn events labelled ("NRCrn") and a total number of left turn events labelled ("NLCrn"). As described above, each acceleration event, $AccXPos_i$,is assigned 0, 1 or 2 depending on the range in which the event falls. For example, an acceleration event of 2.83 m/s^2 is considered "Soft" and assigned a value of two. The assigned values for all these acceleration events are then summed ($\sum_{NAcc} AccXPos_i$). The best possible

Table 4 Scoring system for the driving score

Metric	Range	Score	Level
$AccXPos_i$	$AccX_i > 3.83ms^2$	0	Hard
	$2.83ms^2 < AccX_i \leq 3.83ms^2$	1	Medium
	$0 < AccX_i \leq 2.83ms^2$	2	Soft
$AccXNeg_i$	$AccX_i < -3.73ms^2$	0	Hard
	$-2.73ms^2 \leq AccX_i \leq -3.73ms^2$	1	Medium
	$-2.73ms^2 < AccX_i < 0$	2	Soft
$AccYNeg_i$	$AccY_i < -3.75ms^2$	0	Hard
	$-3.75ms^2 \leq AccY_i < -1.875ms^2$	1	Medium
	$-1.875ms^2 \leq AccY_i \leq 0$	2	Soft
$AccYPos_i$	$AccY_i > 3.75ms^2$	0	Hard
	$1.875ms^2 < AccY_i \leq 3.75ms^2$	1	Medium
	$0 < AccY_i \leq 1.875ms^2$	2	Soft

acceleration score will be the case in which all acceleration events are soft ($2 \times NAcc$). For a trip containing 10 acceleration events, the best possible score a driver can obtain would be 20 if all the acceleration events are soft. The same process is repeated for braking, right cornering, and left cornering.

In the case of speeding, the driver's road speed is compared with the road's posted speed limit using the GPS capability of the G09 device. Since data were recorded on roads within the United States, speed is communicated in terms of miles per hour. Within a trip, the telematic device records the number of times the driver was found speeding (*SpdFreq*) (if speed >8 mph over the posted speed limit) and the distance spent speeding. A speeding score is then the length of the trip not spent speeding divided by the total trip length, as shown in Eq. (5).

$$Speed = \frac{L_S^{'}}{L} \times 10 \tag{5}$$

where $L_S^{'}$ is the length of a trip not spent speeding and L is the length of a trip.

Finally, the driving score is the weighted average of the individual scores as shown in Eq. (6).

$$\begin{aligned} Driving\ Score = {} & 0.3(Speed) + 0.2(Accel) + 0.2(Brake) \\ & + 0.15(Right) + 0.15(Left) \end{aligned} \tag{6}$$

This type of weighted score was developed to: (1) be easy for the drivers to understand and (2) give more weight to metrics that are contributors to road traffic accidents. In addition to seeing a driving score, drivers see a breakdown of their scores on a radar plot. An example is shown in the example.

4 Results

Table 5 shows, for each driver, the trip driving score along with their self-reported mood pre- and posttrip. From Table 5, the data in the column titled "Mood 1" represents their pretrip moods, and the data in the column titled "Mood 2" represents their posttrip moods. The possible moods were "C" – calm; "F" – fatigued; "H" – happy; and "S" – stressed. For ten drivers, 48 trips of complete data were recorded.

4.1 EDA Data

Although a survey can gain some insight into the self-reported emotional states of drivers, it becomes tedious for drivers to complete a survey prior to each driving trip. Their EDA attempts were recorded to gain insight into their physiological

Table 5 The driving score, self-reported mood, and descriptive EDA data for each trip

Driver	TripID	Driving Score	Mood 1	Mood 2	EDA				
					Mean	Median	Std Dev	Max	Skew
CB	1	9.87	C	C	1.219	1.16	0.308	2.4	0.897
CB	2	9.44	H	C	2.723	2.75	0.525	4.626	−0.189
CB	3	10	F	C	0.247	0.231	0.0716	0.886	0.378
CB	4	9.81	S	C	1.36	1.34	0.481	3.477	0.118
DH	1	8.53	C	H	2.305	1.97	0.739	4.936	0.722
DH	2	8.73	H	H	0.279	0.264	0.0627	0.437	0.834
DH	3	9.2	H	C	0.333	0.293	0.108	0.64	1.04
DH	4	8.91	C	S	0.446	0.435	0.0769	0.733	0.447
DH	5	8.63	S	C	0.762	0.766	0.123	1.105	−0.0195
DK	1	9.05	F	F	0.324	0.276	0.192	1.165	1.22
DK	2	8.72	H	H	0.671	0.54	0.498	2.554	1.15
DK	3	8.62	H	H	2.336	2.4	0.805	4.366	−0.534
DK	4	9.4	C	S	2.141	2.29	0.753	4.251	−1.085
DK	5	9.13	C	C	0.0726	0.0794	0.0369	0.407	−0.232
GT	1	8.46	S	S	1.682	1.64	0.408	3.506	0.648
GT	2	8.71	H	H	1.101	0.988	0.441	2.43	0.388
GT	3	8.72	C	C	0.186	0.182	0.0857	0.565	0.767
GT	4	8.2	S	S	0.0692	0.0653	0.0291	0.154	0.496
GT	5	8.55	C	C	0.0365	0.0359	0.00681	0.0807	1.1
WW	1	8.55	C	C	0.54	0.555	0.174	1.34	0.391
WW	2	8.16	C	H	0.49	0.464	0.102	0.782	0.628
WW	3	9.02	C	C	0.204	0.15	0.0931	1.144	1.99
WW	4	8.6	C	C	0.298	0.289	0.0243	0.556	2.63

(continued)

Table 5 (continued)

Driver	TripID	Driving Score	Mood 1	Mood 2	EDA				
					Mean	Median	Std Dev	Max	Skew
WW	5	8.87	C	S	0.378	0.371	0.0558	0.84	2.27
JL	1	8.2	C	C	0.495	0.507	0.121	1.617	−2.2
JL	2	8.35	S	H	0.376	0.406	0.146	1.161	−0.975
JL	3	8.57	H	C	0.354	0.37	0.0865	1.006	−2.82
JL	4	8.5	H	H	0.211	0.25	0.0985	0.738	−1.1
JV	1	8.34	C	C	2.997	2.86	0.687	5.759	0.902
JV	2	9.11	C	C	1.071	0.836	0.79	3.376	0.55
JV	3	8.37	C	C	0.538	1.66	0.473	2.584	−0.24
JV	4	8.51	C	C	0.614	0.151	0.662	2.438	0.698
BB	1	8.11	F	S	0.11	0.11	0.0167	0.176	0.101
BB	2	8.67	H	C	0.577	0.533	0.187	1.172	0.389
BB	3	9.06	H	C	0.0887	0.0923	0.0153	0.119	−0.755
BB	4	8.76	S	S	0.072	0.0769	0.0158	0.104	−0.652
BB	5	9.23	F	C	0.221	0.127	0.183	1.925	1.58
WM	1	8.47	C	C	0.189	0.19	0.0119	0.229	−1.5
WM	2	8.43	C	F	0.277	0.285	0.107	0.602	0.039
WM	3	8.18	S	C	0.771	0.391	1.294	8.836	3.96
WM	4	8.62	F	F	0.173	0.177	0.0511	0.612	3.05
WM	5	8.63	C	C	0.185	0.188	0.0285	0.671	1.04
AOE	1	8.66	C	H	0.118	0.122	0.0263	0.2	−0.913
AOE	2	8.43	H	H	0.0867	0.091	0.0332	0.204	−0.397
AOE	3	8.47	F	C	0.0705	0.0769	0.0381	0.702	1.25
AOE	4	8.57	H	H	0.0537	0.0423	0.046	0.326	1.43
AOE	5	9.25	H	H	0.16	0.112	0.125	0.467	1.02

states while driving. From Fig. 2, the EDA varies by driver and by trip. The driver CB stated being calm for all four recorded trips. However, Fig. 2 shows considerable variability in the distribution of the driver's recorded EDA. The EDA distribution was lowest for Trip3 and highest for Trip2. Interestingly, CB's driving score was highest during Trip3 and lowest during Trip2. Sample boxplots for participants DH and DK are also shown in Fig. 2.

Overall, the average driving score was 8.75, median = 8.63, standard deviation = 0.445, and interquartile range = 0.565. When drivers reported being stressed prior to driving, the average score = 8.67; if the drivers were calm, the average driving score = 8.72. This shows some preliminary evidence that when drivers are "Stressed" prior to driving, their scores are lower than when they are "Calm". For drivers who reported feeling stressed while driving (Mood 2 = "S"), the average driving score = 8.67, and if they reported feeling calm while driving (Mood 2 = "C"), the average driving score = 8.85. Again, this suggests that drivers who reported feeling calm exhibited more eco-driving habits.

Figure 3 shows the distribution of driving scores by Mood 2. For drivers in the "C" state, the data are skewed towards higher driving scores meaning that most of the driving scores are toward the left of the mean. When drivers self-reported being stressed, "S", the upper quantile is smaller than the lower quantile, and the data are skewed to the left, with lower driving scores pulling the mean to less than the median. This suggests that a driving score of 10 is more likely when the drivers are calm. Additionally, the median of the driving scores in the "C" state is lower than when drivers reported being stressed.

Figure 4 shows a plot of the driving score and the average EDA, and Table 6 shows the correlations of the driving score with the EDA descriptive metrics. Interestingly, all correlations are positive, but they are all small. The correlations between the driving score and sdEDA, the driving score and MaxEDA, and the driving score and Skew are all close to zero, indicating no relationship. The correlation between the driving score and mean EDA is greater but still very small. Therefore, given the data, there is no statistically significant evidence to show that when the EDA descriptive statistics increase, the driving score increases.

4.2 *Observations by Driver*

All drivers are unique, and thus, their EDA activities vary. It is difficult to assign an EDA value or range that identifies a "stressed" state for all drivers. For this reason, driving performance is observed for each driver. Table 7 provides a brief description of the observations by driver. The analysis of driver experiences and behaviours during specific trips, considering driving scores, EDA, and reported moods, reveals intriguing patterns. Notably, drivers who achieved their best driving scores often exhibited EDA characteristics aligned with their reported emotional states. For instance, drivers with calm moods tended to record lower average EDA, while those with stress reported higher EDA. Interestingly, the direction of

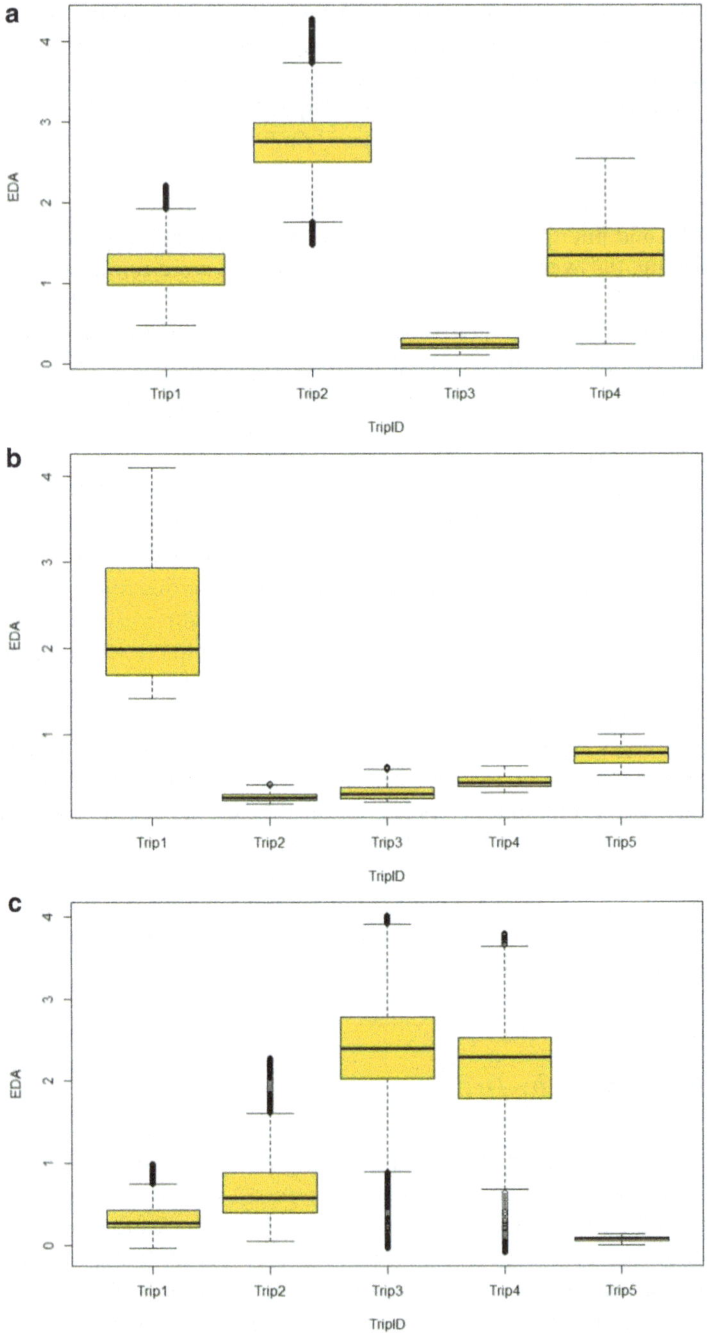

Fig. 2 Boxplots of the EDA distribution for three drivers: (**a**) CB, (**b**) DH, and (**c**) DK

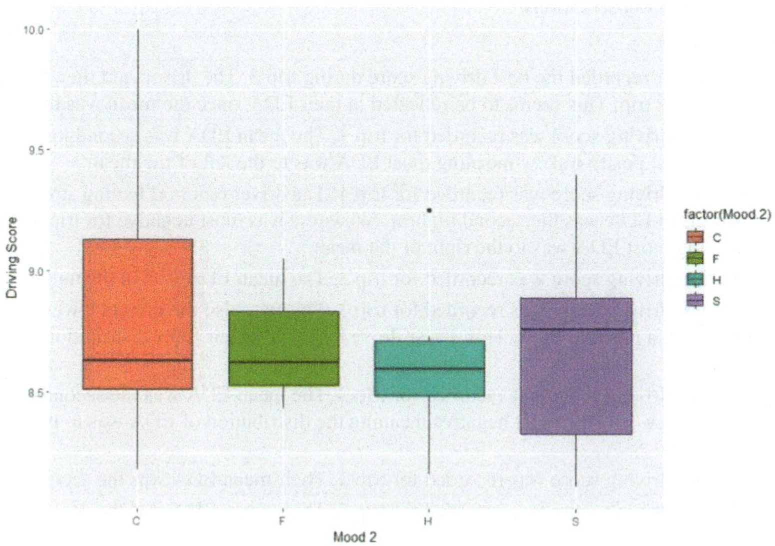

Fig. 3 Distribution of driving scores by mood 2

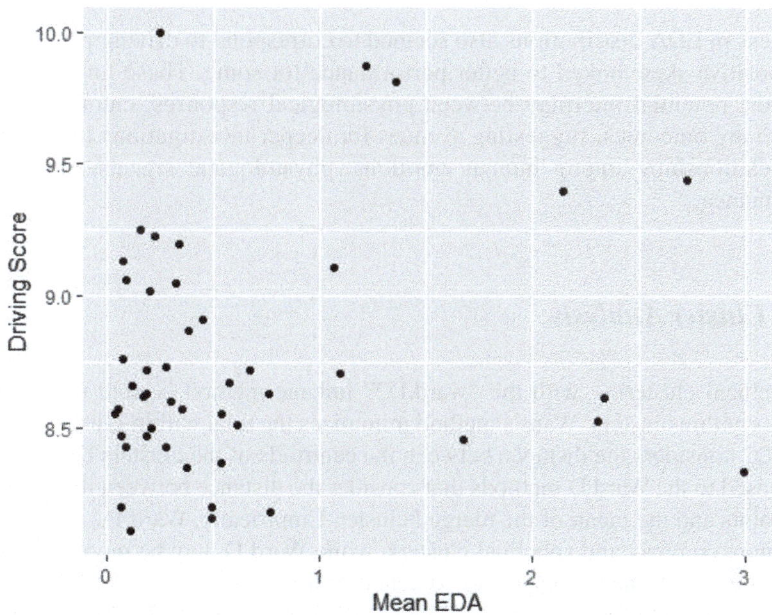

Fig. 4 Driving Score and mean EDA for all trips

Table 6 Correlation matrix

	Mean EDA	Sd EDA	Med EDA	Max EDA	Skew EDA
Driving score	0.1607	0.0318	0.1493	0.0377	0.0366

Table 7 Intra-driver observations

Driver	Comments
CB	This driver recorded the best driving score during trip 3. The driver said they felt calm during this trip. This seems to be reflected in their EDA since the mean was the lowest.
DH	The best driving score was recorded for trip 3. The mean EDA was second lowest with the greatest positive skew meaning most EDA was to the left of the mean.
DK	The best driving score was recorded for trip 4. The driver reported feeling stressed, and their mean EDA was the second highest. Skewness was most negative for trip 4, meaning most EDA was to the right of the mean.
GT	The best driving score was recorded for trip 3. The mean EDA was in the middle.
WW	The best driving score was recorded for trip 3. This was also the drivers lowest average EDA with a positive skew. This driver drove best when their EDA distribution was lowest.
JL	The best driving score was recorded for trip 3. The mean EDA was the second lowest, but the skew was the most negative meaning the distribution of EDA was to the right of the mean.
JV	The best driving score was recorded for trip 2. Their mean EDA was the second highest.
BB	The best driving score was recorded for trip 5. Their mean EDA was the second highest during this trip.
WM	The best driving score was recorded for trip 5. Their mean EDA was the second lowest.
AOE	The best driving score was recorded for trip 5. Their mean EDA was the second highest.

skewness in EDA distributions also seemed to correspond to driving performance, with positive skew linked to better performance for some. These findings underscore the potential interplay between physiological responses, emotional states, and driving outcomes, suggesting avenues for deeper investigations into the complex relationships among human emotions, physiological signals, and driving performance.

4.3 Cluster Analysis

Hierarchical clustering with the "ward.D2" linkage method is used to search for patterns within clusters. Ward's method minimizes the total within-cluster variance. Ward D2 considers the distance between the centroids of the clusters being merged as opposed to the Ward D methods that consider the distance between the individual data points and the mean of the merged cluster. Empirically, Ward D2 tends to produce more compact and spherical clusters, while Ward D may be more sensitive to outliers. Figure 5 shows the generated dendrogram.

From Fig. 6, the Mean EDA and Median EDA show separable clusters. There is a difference in the mean EDA and median EDA between clusters. Cluster 2 has the highest mean and median EDA distribution. The median of the driving scores in Cluster 2 is very close to Cluster 3. However, because of the shape of the boxplot, many of the driving scores fall to the right of the median. Cluster 1 seems to have the "best" distribution of driving scores, in which the scores are generally higher

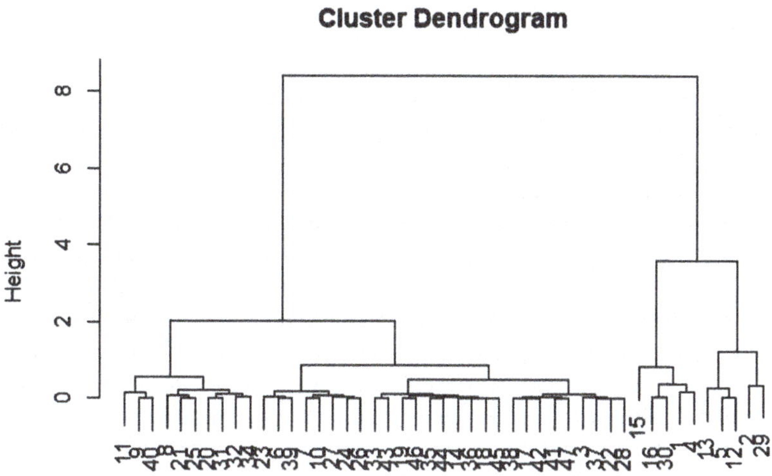

Fig. 5 Dendrogram from cluster analysis

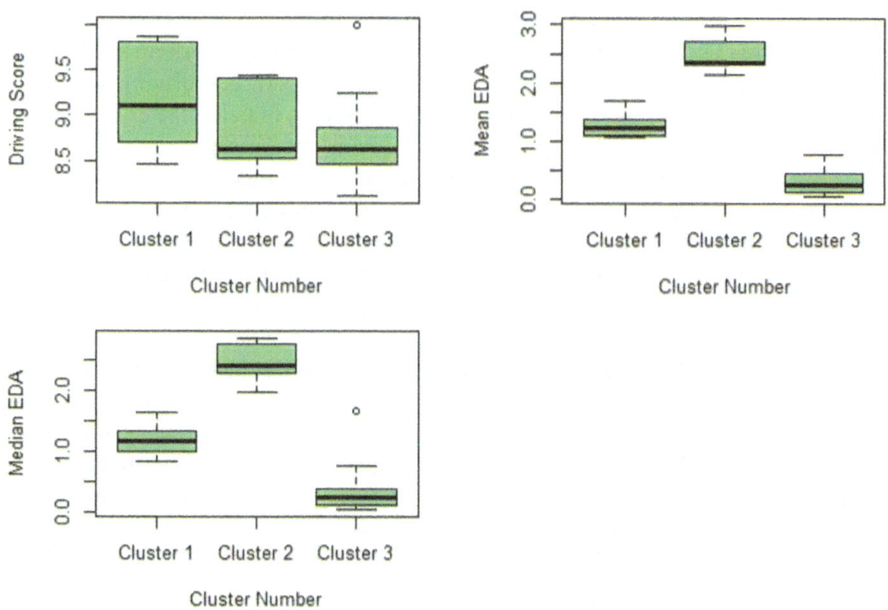

Fig. 6 For each cluster, the above boxplots compare the driving score, mean EDA, and median EDA

than those of the other two clusters. Interestingly, the EDA values are not the lowest and not the highest. This indicates some evidence that driving scores are best when the drivers' EDA is not low but not too high.

5 Discussion

This paper presented a pilot study to investigate the relationship between EDA and eco-driving performance. For fleet managers, stress-related issues can results in increased costs for companies due to accidents, increased downtime, and higher rates of absenteeism. By addressing drivers' stress levels, fleet managers can mitigate these financial burdens. Driver stress can negatively impact the physical and mental health of employees. Chronic stress can lead to various health issues, including hypertension, anxiety, and depression. Caring for drivers' well-being fosters a healthier and more motivated workforce. Stressed drivers are more likely to violate traffic laws and regulations, potentially leading to legal consequences and penalties for the company. A supportive work environment that prioritizes drivers' well-being can improve employee satisfaction and retention rates. Happy and supported drivers are more likely to stay with the company long-term. Fleet companies are responsible for their drivers' actions on the road. High stress levels may lead to aggressive driving behaviours or customer service issues, which can damage the company's reputation and lead to a loss of clients.

Ten drivers wore an Empatica E4 wristband while they completed 5 trips of at least 5 miles in length. The Geotab G09 telematics device was used to record engine data, including acceleration forward, braking and acceleration side to side. It also has GPS capability to identify when speeding occurs. An eco-driving score was used to measure their level of eco-driving. Lower scores indicate fewer eco-driving behaviours.

The highest observed correlation was between the driving score and mean EDA, but this correlation was not statistically significant. Although positive – higher driving scores indicated higher mean EDA – this correlation was not statistically significant. A cluster analysis was also performed to look for patterns within clusters. The cluster dendrogram shows that three separable clusters can be achieved. From Fig. 6, the driving scores in cluster 1 had the highest mean and distribution towards higher scores than the other two clusters. Interestingly, this cluster contained neither the highest nor lowest mean and median EDA. This indicates that the best driving performance for the 48 recorded trips occurred when the drivers were more emotionally aroused. This finding supports the Yerkes-Dodson Law that performance increases with physiological or mental arousal, but only up to a point. When levels of arousal become too high, performance decreases.

5.1 Study Limitations, Strengths, and Future Work

The strength of this study is that it uses naturalistic driving and wearable sensors to observe the eco-driving behaviours of drivers. The E4 sensor and G09 device are both minimally invasive. The preliminary results indicate the need for fleet managers to pay attention to the mental health and stress levels of their drivers. In-vehicle systems to monitor drivers' physiological states while driving.

One of the study limitations is the use of raw EDA data. Another way of analysing skin conductance is to separate it into its phase and tonic components. The phasic component, also known as the skin conductance response (SCR), is a relatively fast variation in skin conductance, while the tonic component, also known as the skin conductance level, reflects slow variation (Benedek & Kaernbach, 2010; Imtiaz et al., 2020). In this study, the phasic component is more significant, as the participant could experience abrupt situations, e.g., sudden braking and sudden accidents. A future study will decompose the EDA signal into its phasic and tonic components and analyse eco-driving performance as the drivers' phasic component changes. A "true baseline", which is the driver's EDA during a calm emotional state, was not recorded in this study. For a future study that uses phasic data, a true baseline is required. In addition to the small sample size, this study limits the sample to young drivers. Future studies can investigate whether similar patterns are observed in different age groups and for a larger sample size.

Vehicle emissions are a major contributor to greenhouse gas (GHG) emissions worldwide. In 2021, the transportation sector was the largest source of GHG emissions in the United States (U.S.) (United States Environmental Protection Agency, 2022). A future study can investigate the relationship between driver's emotional state and their decisions toward sustainable transportation.

Understanding driver stress empowers asset managers to create safer, more efficient, and driver-centric operations. By integrating stress awareness into asset management practices, the transportation industry can achieve higher levels of performance, safety, and driver satisfaction.

6 Conclusion

In this work, the relationship between driver emotional arousal and eco-driving behaviours using naturalistic driving behaviour was investigated. Drivers wore a wristband sensor to record their EDA while driving. An eco-driving score was built using engine data recorded using a telematic device plugged into the OBD II port of drivers' personal vehicle. This pilot study recorded 48 trips of five miles in length from 10 drivers. The results follow the Yerkes-Dodson law. The drivers' best driving scores were observed when they were emotionally aroused but not as the highest level. These results point to the possibility that attention must be given to the emotional state of drivers before they drive. Future work will increase the sample size, incorporate different routes, increase the age range.

Ethics Approval This study received approval from the Institutional Review Board (IRB) with study code STUDY00015895.

References

Ahmed, J., Ward, N., Otto, J., & McMahill, A. (2022). How does emotional intelligence predict driving behaviors among non-commercial drivers? *Transportation Research Part F: Traffic Psychology and Behaviour, 85*(February), 38–46. https://doi.org/10.1016/j.trf.2021.12.013

Alavi, S. S., Mohammadi, M. R., Souri, H., Kalhori, S. M., Jannatifard, F., & Sepahbodi, G. (2017). Personality, driving behavior and mental disorders factors as predictors of road traffic accidents based on logistic regression. *Iranian Journal of Medical Sciences, 42*(1), 24–31.

Al-Mekhlafi, A.-B. A., Isha, A. S. N., & Naji, G. M. A. (2020). The relationship between fatigue and driving performance: A review and directions for future research. *Journal of Critical Reviews, 7*(14). https://doi.org/10.31838/jcr.07.14.24

Andrieu, C., & Pierre, G. S. (2012). Using statistical models to characterize eco-driving style with an aggregated indicator. In *IEEE intelligent vehicles symposium* (pp. 63–68). IEEE. https://doi.org/10.1109/IVS.2012.6232197

Baddeley, A. D. (2000). Selective attention and performance in dangerous environments. *Journal of Human Performance in Extreme Environments, 5*(1). https://doi.org/10.7771/2327-2937.1010

Barkenbus, J. (2010). Eco-driving: An overlooked climate change initiative. *Energy Policy, 38*(2), 762–769. https://doi.org/10.1016/j.enpol.2009.10.021

Beanland, V., Sellbom, M., & Johnson, A. K. (2014). Personality domains and traits that predict self-reported aberrant driving Behaviours in a southeastern US University sample. *Accident Analysis & Prevention, 72*(November), 184–192. https://doi.org/10.1016/j.aap.2014.06.023

Beck, K. H., Ali, B., & Daughters, S. B. (2014). Distress tolerance as a predictor of risky and aggressive driving. *Traffic Injury Prevention, 15*(4), 349–354. https://doi.org/10.1080/15389588.2013.829569

Benedek, M., & Kaernbach, C. (2010). A continuous measure of phasic electrodermal activity. *Journal of Neuroscience Methods, 190*(1), 80–91. https://doi.org/10.1016/j.jneumeth.2010.04.028

Beusen, B., Broekx, S., Denys, T., Beckx, C., Degraeuwe, B., Gijsbers, M., Scheepers, K., Govaerts, L., Torfs, R., & Panis, L. I. (2009). Using on-board logging devices to study the longer-term impact of an eco-driving course. *Transportation Research Part D: Transport and Environment, 14*(7), 514–520. https://doi.org/10.1016/j.trd.2009.05.009

Boodlal, L., & Chiang, K. (2014). *Study of the impact of a telematics system on safe and fuel-efficient driving in trucks*. https://doi.org/10.21949/1502956

Britt, T. W., & Garrity, M. J. (2006). Attributions and personality as predictors of the road rage response. *British Journal of Social Psychology, 45*(1), 127–147. https://doi.org/10.1348/014466605X41355

Broughall, M. (2020, November 5). *What is g-force and how is it related to harsh driving?* GEOTAB. https://www.geotab.com/blog/what-is-g-force/#:~:text=Harsh%20acceleration%20and%20harsh%20braking,-The%20harsh%20acceleration&text=The%20least%20sensitive%20setting%20in,km%2Fh%20in%203.95%20seconds

Butler, G. (1993). Stress Management in General Practice. *Occasional Paper (Royal College of General Practitioners), 61*(August), iv–vi. 1–42.

Cohen, S., Kamarck, T., & Mermelstein, R. (1983). A global measure of perceived stress. *Journal of Health and Social Behavior, 24*(4), 385. https://doi.org/10.2307/2136404

Dehzangi, O., & Williams, C. (2015). Towards multi-modal wearable driver monitoring: Impact of road condition on driver distraction. In *2015 IEEE 12th international conference on wearable and implantable body sensor networks (BSN)* (pp. 1–6). IEEE. https://doi.org/10.1109/BSN.2015.7299408

Eboli, L., Mazzulla, G., & Pungillo, G. (2017). The influence of physical and emotional factors on driving style of car drivers: A survey design. *Travel Behaviour and Society, 7*(April), 43–51. https://doi.org/10.1016/j.tbs.2017.02.001

Fafoutellis, P., Mantouka, E. G., & Vlahogianni, E. I. (2020). Eco-driving and its impacts on fuel efficiency: An overview of technologies and data-driven methods. *Sustainability, 13*(1), 226. https://doi.org/10.3390/su13010226

Gjoreski, M., Gjoreski, H., Luštrek, M., & Gams, M. (2016). Continuous Stress Detection Using a Wrist Device: In Laboratory and Real Life. In *Proceedings of the 2016 ACM international joint conference on pervasive and ubiquitous computing: Adjunct* (pp. 1185–1193). ACM. https://doi.org/10.1145/2968219.2968306

Hassib, M., Braun, M., Pfleging, B., & Alt, F. (2019). Detecting and influencing driver emotions using psycho-physiological sensors and ambient light. In D. Lamas, F. Loizides, L. Nacke, H. Petrie, M. Winckler, & P. Zaphiris (Eds.), *Human-computer interaction – INTERACT 2019* (Vol. 11746, pp. 721–742. Lecture Notes in Computer Science). Springer International Publishing. https://doi.org/10.1007/978-3-030-29381-9_43

Healey, J. A., & Picard, R. W. (2005). Detecting stress during real-world driving tasks using physiological sensors. *IEEE Transactions on Intelligent Transportation Systems, 6*(2), 156–166. https://doi.org/10.1109/TITS.2005.848368

Hellhammer, D. H., Wüst, S., & Kudielka, B. M. (2009). Salivary cortisol as a biomarker in stress research. *Psychoneuroendocrinology, 34*(2), 163–171. https://doi.org/10.1016/j.psyneuen.2008.10.026

Hermawati, S., & Lawson, G. (2019). Identifying the role of human factors in industry 4.0 revolution. In *Contemporary Ergonomics and Human Factors*. https://publications.ergonomics.org.uk/publications/category/contemporary-ehf-2019

Imtiaz, D., Anwar, Y., & Khan, A. (2020). Wearable sensors and a multisensory music and reminiscence therapies application: To help reduce behavioral and psychological symptoms in person with dementia. *Smart Health, 18*(November), 100140. https://doi.org/10.1016/j.smhl.2020.100140

Lee, Y.-C., & Winston, F. K. (2016). Stress induction techniques in a driving simulator and reactions from newly licensed drivers. *Transportation Research Part F: Traffic Psychology and Behaviour, 42*(October), 44–55. https://doi.org/10.1016/j.trf.2016.06.019

linxup. (n.d.). *How do you determine harsh braking?* MOTOsafety Help Center. https://www.motosafety.com/help/alerts-harsh_braking.html. Accessed 20 Nov 2022.

Meng, F., Li, S., Cao, L., Li, M., Peng, Q., Wang, C., & Zhang, W. (2015). Driving fatigue in professional drivers: A survey of truck and taxi drivers. *Traffic Injury Prevention, 16*(5), 474–483. https://doi.org/10.1080/15389588.2014.973945

Mishra, V., Pope, G., Lord, S., Lewia, S., Lowens, B., Caine, K., Sen, S., Halter, R., & Kotz, D. (2018). The case for a commodity hardware solution for stress detection. In *Proceedings of the 2018 ACM international joint conference and 2018 international symposium on pervasive and ubiquitous computing and wearable computers* (pp. 1717–1728). ACM. https://doi.org/10.1145/3267305.3267538

Mishra, V., Sen, S., Chen, G., Hao, T., Rogers, J., Chen, C.-H., & Kotz, D. (2020). Evaluating the reproducibility of physiological stress detection models. *Proceedings of the ACM on Interactive, Mobile, Wearable and Ubiquitous Technologies, 4*(4), 1–29. https://doi.org/10.1145/3432220

Office of Traffic Safety. (2021). *Minnesota motor vehicle crash facts 2020*. Minnesota Department of Public Safety.

Paschalidis, E., Choudhury, C. F., & Hess, S. (2019). Combining driving simulator and physiological sensor data in a latent variable model to incorporate the effect of stress in car-following behaviour. *Analytic Methods in Accident Research, 22*(June), 100089. https://doi.org/10.1016/j.amar.2019.02.001

Romero, D., Stahre, J., Wuest, T., Noran, O., Bernus, P., Fast-Berglund, A., & Gorecky, D. (2016). Towards an operator 4.0 typology: A human-centric perspective on the fourth industrial revolution technologies. In *Proceedings of the international conference on computers and industrial engineering (CIE46)*, Tianjin, China.

Russell, J. A. (1980). A Circumplex model of affect. *Journal of Personality and Social Psychology, 39*(6), 1161–1178. https://doi.org/10.1037/h0077714

Seecharan, T. S. (2021). Encouraging fuel-efficient commuting in young drivers in Duluth, MN. In A. C. Márquez, D. Komljenovic, & J. Amadi-Echendu (Eds.), *14th WCEAM proceedings* (pp. 55–67. Lecture Notes in Mechanical Engineering.). Springer International Publishing. https://doi.org/10.1007/978-3-030-64228-0_6

Seecharan, T. (2022). A framework for gamification to encourage environmentally friendly driving habits. In J. O. P. Pinto, M. L. M. Kimpara, R. R. Reis, T. Seecharan, B. R. Upadhyaya, & J. Amadi-Echendu (Eds.), *15th WCEAM proceedings* (pp. 69–78. Lecture Notes in Mechanical Engineering.). Springer International Publishing. https://doi.org/10.1007/978-3-030-96794-9_7

Stuster, J. (2004). *Aggressive driving enforcement: Evaluation of two demonstration programs*. DOT HS 809 707. National Highway Traffic Safety Administration. https://one.nhtsa.gov/people/injury/research/aggdrivingenf/pages/introduction.html#:~:text=NHTSA%20defines%20aggressive%20driving%20as,gesticulating%2C%20is%20a%20criminal%20offense

United States Environmental Protection Agency. (2022). *Inventory of U.S. greenhouse gas emissions and sinks*. April 14, 2022. https://www.epa.gov/ghgemissions/inventory-us-greenhouse-gas-emissions-and-sinks

Vallès-Català, T., Pedret, A., Ribes, D., Medina, D., & Traveria, M. (2021). Effects of stress on performance during highly demanding tasks in student pilots. *The International Journal of Aerospace Psychology, 31*(1), 43–55. https://doi.org/10.1080/24721840.2020.1841564

Yellman, M. A., & Sauber-Schatz, E. K. (2022). Motor vehicle crash deaths – United States and 28 other high-income countries, 2015 and 2019. Weekly Report. Morbidity and Mortality Weekly Report. Centers for Disease Control and Prevention. https://www.cdc.gov/mmwr/volumes/71/wr/pdfs/mm7126a1-h.pdf.

Yerkes, R. M., & Dodson, J. D. (1908). The relations of strength of stimulus to rapidity of habit-formation. *Journal of Comparative Neurology and Psychology, 18*, 459–482.

Zarkadoula, M., Zoidis, G., & Tritopoulou, E. (2007). Training urban bus drivers to promote smart driving: A note on a Greek eco-driving pilot program. *Transportation Research Part D: Transport and Environment, 12*(6), 449–451. https://doi.org/10.1016/j.trd.2007.05.002

Turuna S. Seecharan obtained her MSc and PhD degrees in Systems Design Engineering from the University of Waterloo, Canada, specializing in "Robust Design Optimization using Metamodels". She is a licensed professional engineer with the Professional Engineers of Ontario and a board member of the International Society of Engineering Asset Management. Presently, she is an Assistant Professor in the Department of Mechanical and Industrial Engineering at the University of Minnesota Duluth. Her current research interests are Gamification and the role of Human Factors in Asset Management.

SPRINGER NATURE

GPSR Compliance

The European Union's (EU) General Product Safety Regulation (GPSR) is a set of rules that requires consumer products to be safe and our obligations to ensure this.

If you have any concerns about our products, you can contact us on ProductSafety@springernature.com

In case Publisher is established outside the EU, the EU authorized representative is:

Springer Nature Customer Service Center GmbH
Europaplatz 3
69115 Heidelberg, Germany

The manufacturer's authorised representative in the EU is Springer Nature Customer Service Centre GmbH, Europaplatz 3, 69115 Heidelberg, Germany. If you have any concerns regarding our products, please contact ProductSafety@springernature.com

Printed and bound by CPI Group (UK) Ltd, Croydon, CR0 4YY
30/09/2026
02225318-0006